CLARK
D0699285

MASQUERADE

The Amazing Camouflage Deceptions of World War II

SEYMOUR REIT

HAWTHORN BOOKS, INC.
Publishers/NEW YORK
A Howard & Wyndham Company

To Irv Leveton, Phil Dempsey, Wayne Booth,
Hal Thresher, Jim Warburton,
and all the others . . .

MASQUERADE: The Amazing Camouflage Deceptions of World War II

 All inquiries should be addressed to Hawthorn Books, Inc., 260 Madison Avenue, New York, New York 10016. This book was manufactured in the United States of America and published simultaneously in Canada by Prentice-Hall of Canada, Limited, 1870 Birchmount Road, Scarborough, Ontario.

Library of Congress Catalog Card Number: 77–70122

ISBN: 0–8015–4931–0

1 2 3 4 5 6 7 8 9 10

Contents

Preface

The Old Testament (Judges 4:12) records one of the earliest of military ambushes. At that time, in about 1125 B.C., a powerful army of Canaanites with "nine hundred chariots of iron," led by Sisera, was preparing to attack Israel. Counseled by the warrior-priestess Deborah, ten thousand Israelites, commanded by Barak of Kedesh, hid themselves on the craggy slopes of Mount Tabor which loomed above the river Kishon. With a small token force, Deborah lured Sisera's army along the river bank toward a narrow marshy plain where the heavy chariots would have difficulty maneuvering. Here Barak's soldiers surprised them, pouring down suddenly from Tabor's slopes and defeating the Canaanite legions.

Homer's *Iliad* tells of the celebrated ruse that supposedly enabled a Greek army under Agamemnon to capture the city of Troy. For nine long years the Greeks had ravaged the countryside and lain siege to Troy, but were unable to pierce the city's massive walls. Finally, the besiegers built a giant horse of wood, ostensibly for use in a religious rite; but the horse was hollow, and concealed inside was a squad of Greek warriors—the first of history's commandos. Leaving the wooden idol behind them, Agamemnon's troops soon embarked on their ships and apparently set sail for home. Overjoyed, the Trojans—despite warnings from the seer, Cassandra—hauled the great horse inside the walls as a trophy of war. Late that night the Greek legionnaires crept out of hiding, threw open the gates in the wall, and Agamemnon's battalions, waiting outside in the darkness, poured through the breech to overwhelm the defenders.

Centuries later in another part of the world, a British force under Major John Pitcairn fought the first skirmish of the American Revolution at Concord, Massachusetts. Following that brief battle in April 1775, Pitcairn marched his troops back toward Boston, and during the fifteen-mile withdrawal his men were constantly harassed by colonial militia. George III's regulars, formed in impeccable ranks, resplendently decked out in scarlet coats, white crossbelts, and white breeches, were gaudy targets. As the history books all note, the colonists in their buckskin and drab farm clothing practiced guerrilla tactics, filtering silently through the New England woods, firing from behind trees, hedges, and stone walls; and by the time the weary redcoats reached their Boston garrison they had had sustained many casualties.

Human resourcefulness is no substitute for firepower, but in warfare it can significantly alter the results; and military records are replete with accounts of successful ambushes, tricks, traps, and hoaxes.

The particular camouflage deceptions described in this book all took place between May 1940, when the Germans first overran western Europe, and September 1945, when the Japanese formally surrendered on board the U.S.S. *Missouri* in Tokyo Bay. During the turbulent years of World War II, the tactical use by both sides of visual and aural subterfuge reached new levels of skill and acuity; and at no other time and in no other war have the deceptive arts played so subtly effective a role.

Given today's push-button technologies and computerized weapons of mass destruction, it is unlikely that the remarkable human episodes recorded in these pages will ever transpire again.

(Editor's note: A glossary of military terms will be found on page 225. Author's notes and sources begin on page 231.)

Acknowledgments

During the past few years, many people generously extended their time and effort in helping me to prepare this book. In addition to those directly interviewed and quoted herein—and to whom I am most grateful—I wish to thank the many others whose assistance was of value. For directing me to the right people and sources during the early months, I particularly must thank Gen. Sir John Hackett, Flt. Lt. Robin A. Brown (RAF), Richard Frost, formerly of the 10th Squadron (RAF), Robert Schulberg, formerly of the 82nd Airborne Division, Carolyn Trager, Lewis Richardson, Barnett Simons, Robert Kaye, and Elin Waite of *Westways* magazine.

For their help with specific research needs and problems, I am indebted to Marilla B. Guptil and David Eggenberger of the U.S. National Archives and Records Service, James Gilreath of the Library of Congress, Hannah Zeidlik and Mary Haynes of the Office of Military History, and Richard Brower's staff in the Microfilm Department of the New York Public Library. Thanks are also due to Vicki DeStefano of the U.S. Army Audio-Visual Agency at the Pentagon, and Dana Bell and Margaret B. Livesay of the USAF 1361st A-V Squadron, in Arlington, Virginia.

For assisting me in exploring and developing the British camouflage story, my appreciation goes to A. J. Bavistock and the staff of the Public Record Office in London, Peter J. Simkins and Michael Willis of the Imperial War Museum, and Peter G. Merton of the Royal Air Force Museum at Hendon.

For their help and cooperation, I am also grateful to William H. Schmidt and Sol London of the Lockheed Aircraft Corporation, Burbank, California, Harris and Margaretta Colt of The Military Bookman, New York, and author David Fisher, who made available to me much useful information on camouflage in North Africa. Additional and most sincere thanks go to Lorin Driggs, my invaluable secretary and official G.L.C.; Judith Sachs, my editor at Hawthorn Books; and finally to the patient and skilled indexer of this volume, to whom I am fortunate enough to be married.

And, after all, what is a lie?
'Tis but the truth in masquerade.

—George Gordon, Lord Byron
(*Don Juan*, Canto XI)

1

Image and Illusion: A Prologue

"Hold out bait; entice the enemy . . ."

The Luftwaffe squadron, nine Junkers 88s, came over the Hampshire coastline at sixteen thousand feet and veered northwest, their goal a large aircraft plant on the outskirts of Bristol. As the bombers swept inland they ran into searchlight probes and sporadic antiaircraft fire, but British defenses this far from London were relatively feeble and the night haze gave them cover. The *Fliegerhauptmann* in the lead plane shaped his course methodically; and now through a spill of moonlight he could see the Avon River glinting on Salisbury Plain, drawing them obligingly toward the target.

Twelve minutes later the navigator leaned forward, tapped him on the shoulder and gestured to the left. The pilot followed his pointing finger and nodded—in the distance he could just make out the luminous bulk of their objective. The British had done an inept blackout job; the sprawling factory was faintly distinguishable in the darkness, and here and there he could even catch a glimmer of blue-shaded headlights from trucks moving near the loading platforms.

The Junkers' bomb bay doors rumbled open and the pilot

turned into his final approach, ignoring the ground batteries which had begun firing at them heavily. Moments later the ship lifted as two tons of explosives headed downward, and behind the leader the other planes of the *Staffel* also unloaded their bombs. As the pilot banked sharply, turning toward the Channel, he glanced over his shoulder. It had been a near-perfect drop; he could see dozens of fiery explosions blanketing the factory area.

The *Fliegerhauptmann*, pleased with the success of the mission, framed the report he would make when they reached their base. He had no way of knowing that his squadron's bombs had landed accurately on twenty-three acres of deserted English countryside, rigged with a complex pattern of "Q" lights which had, by design, been left partially exposed. Truck headlights (mounted on mechanized trestles) and AA gun flashes (set off by remote control) were all part of the charade. The real plant—two miles to the east and meticulously blacked out—had won another night of uninterrupted production, and by morning a few new Spitfires would be ready to join in the Battle of Britain.

As early as 500 B.C. a leading Chinese tactician, Sun Tzu, wrote, "All warfare is based on deception."[1] The military advisor to the Chou emperors went on to counsel:

> When able to attack, we must seem unable; when using our forces, we must seem inactive; when we are near, we must make the enemy believe that we are far away; when far away, we must make him believe we are near. Hold out the bait; entice the enemy. Feign disorder and crush him.

Hoax and deception have always played a role in warfare and Sun's doctrines have been adapted in varying forms by Marlborough, Frederick the Great and Napoleon; by American colonists during the War for Independence; by Meade and the

Army of the Potomac.[2] In World War I, feint and subterfuge were employed brilliantly by Sir Edmund Allenby against the Turks at Gaza and Beersheba, and by the navies of all the belligerent forces. But at no time were military deception and its ally, camouflage, more widely and effectively used than in the campaigns of World War II.

The battles of that war, extending over a six-year period and spreading to all parts of the world, were total and uncompromising; but below the visible crust of the fighting other wars went on—hidden battles involving espionage, sabotage, cryptology, partisan resistance, counterintelligence, psychological manipulation, secret technical experiments, electronic artifice, and deception through conjury and camouflage. The success of these strategies had a significant—sometimes decisive—bearing on the outcomes of the war's bloody campaigns. The planting of a spurious map, for example, led to the first defeat of the Afrika Korps at El Alamein. The faking of a wireless message helped the U.S. Navy win the battle of Midway. The brilliant use of decoys and camouflage duped the Germans and eased the Allied path to Normandy.

In the natural world where camouflage is a survival device, many living creatures depend on protective shape and coloration.[3] Ptarmigan chicks wear mottled feathers which fade into the underbrush. The praying mantis looks like a small green stem or a blade of grass. *Hyla regilla,* the Pacific tree frog, resembles a piece of bark. Tiny insects called tree hoppers look like thorns. The katydid is a walking green leaf. *Phyllopteryx eques,* a tropical sea horse, can pass for a bit of floating kelp. The reddish-brown coat of the fallow deer, spotted with white, helps it to blend with the dappled light of the forest. In winter the brown fur of the weasel miraculously becomes the snowy pelage of the ermine. Many species of fish have countershading to protect them from predators, showing dark brown or green when seen from above and silvery blue tones when viewed from below. And chameleons, shrimps, and squids can change color to match their surroundings.

Camouflage in warfare—the fine art of misdirection—follows similar principles, but has a two-part mission: depending on military needs it can *conceal* from an enemy the true picture, or *mislead* an enemy into accepting a false one. While the fighting—and dying—occupy center stage, the camoufleurs serve in effect as the stagehands, creating and shifting the necessary scenery in an attempt to make the fighting more productive and the dying less indiscriminate.

In World War II, as in earlier conflicts, the aim of camouflage was to delude rather than destroy. The weapons most frequently used for this were paint, canvas and fishnets, lumber and chicken wire, fiberboard, smoke, lights, plaster, papier-mâché, and every form of natural foliage from pine branches to palm fronds. Added to all those prosaic tools were, of course, the essential ingredients of creativity and imagination.

On the battlefield itself the conjurors often had to improvise. In the middle of a Burmese swamp or a Tunisian wadi, paint and chicken wire aren't easy to come by, and the illusionists frequently had to scavenge for materials. Every wrecked plane, burned-out truck, ripped parachute, and battered oil drum was pounced on as a valuable resource and nothing was overlooked. During the desert campaigns in North Africa, sand-colored paint for tanks and half-tracks was desperately needed but often in short supply. At one base the camoufleurs solved this shortage by mixing a large batch of whitewash with a case of Worcestershire Sauce, which had turned up unexpectedly on a beached transport. When the sauce ran out, they found to their great delight that camel dung made a splendid substitute. But in spite of bizarre problems and improvisations, the point of it all remained simple and consistent: *do everything you can to baffle, confuse, and mislead the enemy*.

Among the principal belligerents in the war, camouflage approaches and techniques varied. The American forces, suspicious at first of "passive" cover and concealment, were converted by the attack on Pearl Harbor and became efficient and enthusiastic practitioners, and as the fighting went on they introduced valuable innovations. Their Pacific enemy, the Japanese, were geniuses at individual camouflage and created deadly ruses well suited to jungle fighting. However, in the rear echelons—which involved the protection of factories, rail yards, and supply depots—they were surprisingly naïve and incompetent. Russian camouflage was generally conventional, except for the brilliant use by Soviet divisions of whitewashed armor, snow capes, and similar winter camouflage on sub-zero fronts like those at Tikhvin, Voronezh, and Stalingrad.

The camoufleurs of Germany and Great Britain undoubtedly showed the greatest originality and inventiveness. After the Luftwaffe lost air supremacy—and the Germans were forced gradually on the defensive—they learned to hide their war factories in vast underground caverns. They covered whole sections of Berlin's Tiergarten and the Unter den Linden under huge camouflage nets, to confuse B-17 pilots who used these landmarks for target orientation. They also made excellent use of decoys, dummies, and diversions. During the 1943 fighting at Benghazi, RAF pathfinders often flew over the city at night and dropped flares to guide the Allied bombers. The Germans responded by setting off their own flares in such a way that the planes "lost" the city and frequently attacked deserted sections of the beach.

The postwar U.S. Strategic Bombing Survey, discussing German camouflage, reported, "Protective concealment was practised with greater variety of materials, probably with greater ingenuity, and certainly with greater expenditures of manpower, than had been used by any warring nation previously."[4]

But the Germans sometimes negated their own hoaxes by the very methodical and systematic way they went about them. At Kiel and Bremen in 1942, scores of 500-ton U-boats were under construction—part of Chancellor Adolf Hitler's cherished plan to crush England by cutting her ocean supply lines. Technicians at the shipyards developed an elaborate system of camouflage screens to hide the important work being done, and as progress was made on each new submarine the screens were neatly and carefully extended, step by predictable step. Their Teutonic precision was very helpful; Allied planes photographed the docks regularly, and it was easy for the interpreters studying the aerial prints to know just what stage each vessel was at, and exactly when it would be ready for launching. David Brachi, one of the RAF's photo specialists, remarked at the time, "The Germans are so methodical about their camouflage that once you get to know their methods you can tell quite a lot from the camouflage itself."[5]

Of all the combatants (the U.S. Bombing Survey notwithstanding), no nation matched Great Britain in terms of the scope and daring of its deceptive strategies. Reeling back from the Dunkirk disaster in 1940, the British were desperately short of planes, tanks, guns, ammunition, military supplies—and above all, time. They needed precious time to recover, rearm, hold off the Luftwaffe, and prepare defenses against the Wehrmacht divisions massing on the far shores of the Channel.

In the subsequent Battle of Britain the RAF pilots fought like heroes and their dramatic story has often been told; but on the ground below, other heroes were also at work. The Royal Engineers and the British Home Guard performed prodigies of deception aimed at convincing Hitler that England was stronger and far more prepared than he realized. Pounded continually during the "blitz," the British created a vast web of fake targets which helped to protect their real installations and siphon off the German night attacks. Using little more than ingenuity and the so-called Q lights, they simulated dummy docks, factories, shipyards, fuel and supply depots that attracted thousands of tons of enemy bombs; and in that same period they built numerous decoy airfields, which actually drew *more* Luftwaffe raids than the genuine RAF bases.

England was also prepared for enemy tanks and paratroopers. Fake gun batteries were set up along the coastline to fool German reconnaissance planes and to pad out the thin line of real defenses; while farther inland there were scores of camouflaged strong points, minefields, and hidden explosive devices. On Parliament Square in London, under the unsuspecting noses of daily passersby, an ordinary-looking newsstand concealed an antitank gun. Other gun emplacements masqueraded as old wooden shacks, bridge abutments, tool sheds and innocent shops, and one pillbox was disguised somewhat caustically as a large marble war monument.

Later on when the threat of German invasion faded, the British and Americans continued these phantasms on a larger scale, creating unprecedented tactical and strategic illusions, and proving in the process that camouflage as a combat weapon could be as valuable militarily as a paratroop battalion or an armored brigade. Many of these dramatic episodes have never before been fully reported; a few have been documented earlier, but as largely isolated incidents. The aim of this particular narrative is to weave all of the many disparate strands and little-known threads together, to tell the story in its unique entirety, and to frame it finally in its historic context. Since covert activities such as espionage, counterintelligence, and cryptology have been described elsewhere in detail, those will be covered here only as they relate specifically to camouflage hoaxes and operations.

In the period between the two world wars, Britain's Prime Minister Winston Churchill, a student of military strategy and doubtless a disciple of Sun Tzu, wrote:

> There are many kinds of maneuvers in war, some only of which take place on the battlefield. There are maneuvers far to the flank or rear. There are maneuvers in time, in diplomacy, in mechanics, in psychology; all of which are removed from the battlefield, but react often decisively

> upon it, and the object of all is to find easier ways, other than sheer slaughter, of achieving the main purpose.[6]

In saving lives the use of fraud and deception, even Machiavelli agreed, is worthy and honorable. During World War II, camouflage deception served as a kind of invisible armor for the vulnerable men who fought the battles; and finding in Churchill's words those easier ways of achieving the main purpose was the chief mission of the camoufleurs. At times they failed, at times they achieved remarkable success—and this is their story.

2

Bodyguard—the Houdini Touch

1943–1944

In the early spring of 1944, as slate-gray clouds lifted over the coast of France and the first green buds began to appear in the apple orchards of Normandy, Colonel Alexis Baron von Roenne sat at his elegant desk and brooded.

Von Roenne, a cool and haughty aristocrat in the classic Junker tradition, was chief of *Fremde Heere West,* the arm of the German General Staff which collected and evaluated intelligence data about the western Allies. The normally unruffled colonel was growing increasingly agitated, and with good reason. Invasion weather was coming, which meant the likelihood of a huge cross-Channel attack by Allied armies—and day after day FHW Headquarters at Zossen, south of Berlin, was swamped with questions from Wehrmacht commanders all over Europe.

Where would the Allies make their landings? When would the invasion take place? How many divisions were involved? What type of armor and fire power would the invading troops have? What was their order of battle?

Von Roenne's responsibility, along with other German intelligence units, was to sift through the great mass of reports that flooded his desk and give accurate—or relatively accurate—answers to this growing clamor for facts and information. Actually, FHW already knew a great deal about the Allies' plans. Much of their information came from German secret agents operating in England. Other data came from Wehrmacht cryptographers and wireless intelligence experts who, by 1944, had developed the monitoring and analysis of enemy messages to a remarkably fine art. Still other information came from the Luftwaffe's reconnaissance planes, and from captured members of the resistance in Nazi-occupied territory.

These sources provided endless bits and pieces which could be painstakingly fitted into the puzzle to create a *Feindbild* or "picture of the enemy"—yet there were frustrating gaps that nagged at Alexis von Roenne's meticulous Prussian mind.

Why, for instance, was the British Fourth Army assembling along the Scottish east coast, and what was its precise mission? The bits and pieces were all there, right at his patrician fingertips. He knew that the commander in chief of the British Fourth was General Sir Andrew Thorne, a prestigious officer who had once served as England's military attaché in Berlin. He knew that Fourth Army Headquarters was centered in bunkers deep under Edinburgh Castle, and that there was a Fourth Army corps at Stirling and another at Dundee. He had also discovered that a Russian liaison officer, Klementi Budyenny, had just been assigned to the Edinburgh Command; and working in close support, the U.S. Fifteenth Corps, under General Wade Haislip, was going through full-scale invasion exercises. In all, FHW had pinpointed a strong assault force of over two hundred fifty thousand men, equipped with hundreds of tanks, armored vehicles, field artillery pieces, and its own tactical air command. Two secret agents in Scotland had radioed additional details, down to the design of the Fourth Army's shoulder insignia, which was a blue-and-red square with half a figure eight outlined in gold. Reports from Nazi sources in Scandinavia indicated that this force was preparing for an imminent invasion of Norway. Everything pointed to it. Von Roenne kept systematically filling in the few remaining gaps, and by mid-May he and the German High Command knew an astonishing number of details about General Thorne's troops, their disposition, and the scope of their activities.

What they didn't know—and wouldn't learn until long after—was that there was no British Fourth Army. It simply didn't exist. The army's menacing tanks were made of inflated rubber, its numerous landing barges were built of canvas and chicken wire, its busy wireless traffic was a hoax—and its total military complement consisted of twenty discreet officers, a squad of ingenious radio operators, and a company of hardworking artists and carpenters adept at creating illusions.

"Bulgy" Thorne's army was a phantom, as intangible as the airwaves that buzzed with its bogus messages; but for all its lack of substance, it played a key role in the success of the Allied D-day landings. In effect its contribution was far greater than the equivalent in live troops and real tanks, because the threat to Norway by the "British Fourth" helped to pin down hundreds of thousands of German soldiers who could have been thrown into the beachhead fighting with disastrous consequences to the invaders.

The creation of this imaginary, invisible army was only one small part of a much larger design, a highly complex plan with a very simple aim: to trick the Germans and in the process save Allied lives. The scheme had its roots primarily in the thinking of Winston Churchill, who had always been an advocate of deceptive tactics in modern warfare. Years before, in speculating about World War I, he concluded, "There is required for the composition of a great commander not only massive common sense and reasoning power, but also an element of legerdemain, an original and sinister touch, which leaves the enemy puzzled as well as beaten."[1]

Legerdemain was the operative idea. Earlier in the fighting the Allies had used camouflage and deception with much success, but now the need was more vital than ever. The English Channel was a tricky and dangerous barrier for invading armies to cross, and waiting for them on the far side was Hitler's formidable Atlantic Wall manned by a million trained and heavily armed *Feldgrau*. The planners also knew that Hitler had ordered his most brilliant commander, the legendary Erwin Rommel, to take charge of the Channel fortifications and make them impregnable.

From the Allies' point of view, he couldn't have picked a more dangerous man. For many months, while the fighting went on elsewhere, the Atlantic Wall had been neglected; but Rommel, who took over its defenses in November 1943, began to make sweeping changes. Within a few months the efficient, hard-driving field marshal had managed to install acres of new minefields containing over four million mines.[2] Along the cliffs he added dozens of heavy gun emplacements built of reinforced concrete; below the French coastal waterline he hid thousands of tank and personnel obstacles, many of them wired with hair-trigger explosive devices. Rommel, a camouflage enthusiast, also added dummy minefields, smoke-making apparatus, and decoy gun batteries so realistic that they drew repeated attacks by Allied bombers. Most of this work was done in the Pas de Calais area, at the narrow neck of the Channel, but the Norman and Breton coasts were being strengthened as well.

Reports of these renewed activities quickly reached SHAEF (Supreme Headquarters Allied Expeditionary Forces), adding to the growing pressure on the American and British commanders. The basic Allied plan for the liberation of western Europe had been code-named "Operation Overlord," and the strategy for crossing the Channel and carving a foothold in France was called "Neptune." The planning for Overlord had been going on for almost three years, and as the plans took form so did the pitfalls. "We were gambling heavily," recalled an American staff officer, "and somehow the stakes kept getting higher and higher."

It was clear enough to the Allies that in the long run, with their huge industrial and military potential, a safe bridgehead would sooner or later mean victory. But how could a safe bridgehead be assured? And what if—thinking the unthinkable—the Neptune landings failed? Anthony Cave Brown, the noted military historian, pointed out the calamities that could flow from an Allied defeat:

> As everyone knew, if the invasion failed, all else would fail. Britain would have to seek terms, for she would commit everything in her armory to this attack. The Americans, appalled by the bloodshed and the magnitude of the disaster, would almost certainly reject President Franklin D. Roosevelt when he came up for re-election late in 1944, and seek a victory against Japan before deciding whether to attempt a second invasion. Then Hitler would be able to concentrate his entire might against Russia, with every prospect of defeating the Red Army and emerging as master of Europe.[3]

The German dictator was well aware of the stakes involved. Ten weeks before the invasion he lectured his field commanders, "Once the landing has been defeated it will under no circumstances be repeated by the enemy." He then went on to talk about the "heavy casualties" and the "crushing blow to their morale," which a failure would bring.[4]

At that point in the war, Hitler had reason to be hopeful. Even after months of heavy bombing by the Allies, Nazi production figures remained impressive. In 1943, German factories managed to turn out over 2 million tons of ammunition. They also produced 11,800 tanks, 26,000 pieces of field artillery, and some 20,000 aircraft; and the Wehrmacht, hard-pressed on the Russian front, could still muster some seventy or eighty divisions for use in Norway and France.[5]

Josef Goebbels, Germany's propaganda minister, boasted on Radio Berlin: "We have fortified the coast of Europe from the North Cape to the Mediterranean and installed the deadliest weapons that the twentieth century can produce. That is why any enemy attack, even the most powerful and furious possible to imagine, is bound to fail."[6]

Despite Goebbels's usual hyperbole and bluster, the boast had a disquieting plausibility. An attack by sea against a well-fortified coastline is the most dangerous of all operations, as the Allies knew. In August 1942, Canadian troops led a raid on the French port of Dieppe. Of the six thousand men who took part, less than half returned; and some thirty-six hundred were killed, wounded, or captured. In September of the following year, the U.S. Fifth Army under General Mark Clark made landings at Salerno on the coast of Italy; the invaders met heavy resistance, suffered fifteen thousand casualties, and barely held on until reinforcements arrived. Landing at Anzio in January 1944, the U.S. Sixth Corps under General John Lucas also met disaster; for weeks the entire Corps was trapped on a small exposed beachhead while its men were picked off by the thousands.

For the Neptune planners these were terrible recollections; and added to all that was one central fact: D-day was an operation of such size that *it was virtually unrepeatable.* If Hitler's panzers threw the attackers off the beaches, the Grand Alliance would be paralyzed for years, if not competely shattered. Germany's engineers would have time to complete their new weapons—jet aircraft, long-range submarines, and deadlier rockets to supplant the V-1 and V-2 missiles that were already being prepared for use against London. As for Stalin, normally distrustful of his Western Allies, he might decide that the invasion had been a fraud and try, if possible, to make a compromise peace with the Nazis.

Then there were the men and women of the resistance waiting patiently in occupied Europe—the armed guerrilla units of France, Belgium, Holland, and Scandinavia. For years they had harassed and sabotaged the Nazis and were ready to come out in force as soon as the invasion began; but if they showed themselves openly and the Allies were defeated, they would also be destroyed.

"When you strike at a king," said Ralph Waldo Emerson, "be sure you don't fail." The advice may have occurred to Britain's prime minister, who reportedly had frequent nightmares during this period. The American leaders also felt the pressure. President Roosevelt showed a calm facade in public but fretted endlessly over the invasion plans with his advisors. General George Marshall and the Combined Chiefs of Staff were anxious and worried. General Dwight D. Eisenhower, commanding the Allied Expeditionary Forces, wrote his friend Brehon Somervell, "Tension grows and everybody gets more and more on edge. In this particular venture . . . we are putting the whole works on one number."[7] And Walter Bedell Smith, Eisenhower's chief of staff, perhaps with Dieppe and Salerno in mind, gave Neptune no more than "a fifty-fifty chance."[8]

The planners at Whitehall and the Pentagon knew that something had to be done to improve those odds—something beyond the statistical massing of men, tanks, planes, ships, and weapons. To be successful the invasion (or liberation, as Eisenhower preferred to call it) had to somehow catch the defenders off balance. A special element of cover and surprise was needed—but that was precisely the problem. How could they hide an invasion? How could they achieve surprise with a huge attack involving thousands of vessels, masses of tanks, half a million men, and mountains of ammunition and supplies?

Neptune could hardly be programmed as if it were a small night raid by a squad of commandos with blackened faces. Cover and surprise would have to be gained, if at all, in some very special way. It would take a real act of legerdemain—a deft combination of camouflage and misdirection that would deceive the Nazis as to the Allies' plan and (even more important) keep them deceived for a critical period *after* the invasion was launched. Getting ashore was one thing; staying there was another. The Allies would need time on the beaches to consolidate, reinforce, and bring up supplies, but the German generals planned to strike fast and smash any assault at the high-water mark. "We must stop the enemy in the water,"

Rommel had said, "and destroy his equipment while it is still afloat."[9]

How could the invaders delay this and win an extra period before the panzers counterattacked heavily? It was a critical challenge—and the kind Churchill responded to. With his support and guidance every possibility was explored, and at the Teheran Conference in November 1943, "Operation Bodyguard" was finally activated. Bodyguard (originally called "Plan Jael") came under supervision of the London Controlling Section, directed by Colonel John H. Bevan, and was the code name for cover and deception tactics to be used in the liberation of France. The London Controlling Section, or LCS, was Britain's central agency for covert activity and counterespionage in *all* theaters of war. It had been created at the start of the conflict and its extraordinary history remained a well-guarded secret for thirty years, until the lifting of the Official Secrets Act and other military regulations finally made many of the facts available.

Bevan's LCS had two main operational branches: the OSS, America's secret intelligence service under General William J. Donovan, and MI-6, Britain's secret intelligence service under Admiral Hugh Sinclair, later directed by General Stewart G. Menzies. These men formed the tip of a bizarre and mysterious intelligence iceberg, which encompassed at its great base countless clandestine intrigues in every part of the world.

Working below the iceberg's peak were resourceful deputies such as Kermit Roosevelt, General Colin Gubbins, Major Derrick Morley, Lieutenant Colonel Ronald Wingate, Princess Noor Inayat Khan, Commander Ian Fleming, Duncan Sandys, Allen Dulles, and General Donovan's aide, Dr. Andor Klay. Others included Colonel David Bruce, Dr. William Maddox, Wing Commander Dennis Wheatley, Lester B. Pearson, Hugh Trevor-Roper, Sister Henriette Frede whose convent in Paris doubled as an MI-6 headquarters, and William Stephenson, the legendary figure known as "Intrepid." And below all these came level on hidden level of camoufleurs, cryptographers, spies, radar experts, wireless technicians, sonic engineers, counterespionage agents, cipher analysts, photo interpreters, psychologists, resistance liaison officers, saboteurs, artists, actors, forgers, carpenters, and magicians.

This arcane network of specialists took on scores of difficult, often dangerous assignments with one common purpose: To create, as Colonel Bevan put it, ". . . a war plan that was just close enough to the truth to seem credible to the führer, but would mislead him completely about the time and place of the invasion."[10]

Harry Houdini, the famous escape artist, had a simple formula for performing magic. It was chiefly, he claimed, "the art of making people look somewhere else, when they think they are watching *you*."[11] Now the magicians of Bodyguard would try to apply their own Houdini touch; but instead of a receptive theater audience, they would perform for Hitler, Rommel, Field Marshal Gerd von Rundstedt, the German General Staff, Colonel von Roenne of FHW, General Walter Schellenburg of *Amt Mil*, and the agents of the *Sicherheitsdienst* or SD, the intelligence arm of the Nazi SS.

The Germans, of course, were on their guard. Having been fooled before, they expected the Allies to try new hoaxes and were determined (to the point of paranoia) not to be tricked again. Every deception has its counterpart in detection methods, and the Allies knew that one blunder or false step could wreck the whole carefully built illusion. Lieutenant General Frederick Morgan, SHAEF's deputy chief of staff, later described the problems of the Bodyguard planners:

> The great shadow-boxing match had to go on without a break. One bogus impression in the enemy's mind had to be succeeded by another equally bogus. There had to be an unbroken plausibility to it all, and ever present must be the ultimate aim, which was to arrange that the even-

> tual blow would come where the enemy least expected it, when he least expected it, and with a force altogether outside his calculations.[12]

Operation Bodyguard soon led to numerous associated plans and sub-plans, and with the Allies' penchant for enigmatic labels the conference rooms of the OSS and MI-6 buzzed with code names such as "Glimmer," "Diadem," "Gaff," "Jupiter," "Adoration," "Titanic," "Ironside," and "Zeppelin." But the most ambitious of all these, the hoax that spawned whole phantom armies to mislead von Roenne and the other experts, was known as "Operation Fortitude."

The Fortitude ruse, or set of ruses, supervised by SHAEF's General Harold R. Bull, was described officially as a plan to:

> . . . Cause the Wehrmacht to make faulty troop dispositions by military threats against Norway.
>
> . . . Deceive the enemy as to the correct target date and target area of Operation Neptune.
>
> . . . Induce the enemy to make faulty tactical dispositions during and after Neptune, by threats against the Pas de Calais.[13]

The cold, spartan phrasing of that directive gave little hint of the underlying drama—or the timing, security, and human effort involved. During World War I, infantrymen learned not to peer carelessly over the tops of their trenches; it was much safer, in case of sniper fire, to prop a helmet on a rifle and poke that up first. The Fortitude deception would create a metaphoric helmet for the Allied armies—not merely drawing enemy fire but riveting the Nazis' attention on areas *miles away* from the intended battlefront.

As 1943 drew to a close General Bull, a crusty veteran with the unlikely nickname of "Pinky," sat down with his staff—their faces lined with strain—to plan the huge job confronting them. Later in the war "Pinky" Bull, as Eisenhower's

G-3, would be responsible among other things for riding herd on the irrepressible General George Patton; but that would be child's play compared to what he faced now.

To build their hoax, Bull and his people would use every form of camouflage and decoy, plus wireless deceptions, misleading "messages" to the underground, spurious tips from double agents, carefully planted rumors, counterfeit orders of battle, and diversionary air and naval maneuvers, all involving a network that stretched from Moscow to Reykjavik, from Marseilles to the Ionian Sea. It was the Houdini touch on a global scale, a strategem more complex than anything they had ever tried, and the potential results gave it extra urgency. If the Allies succeeded, their conjuror's war-within-a-war would fool Hitler, save thousands of lives, and perhaps alter the precarious fate of the invasion itself.

But all through the weeks of planning and preparation, as each ingenious element fell secretly into place, one question persisted: Could the scheme be made to work?

3

The Phantoms of Fortitude

1944

On a mild April morning of the fateful invasion year a Home Forces clerk, bicycling to work near Kirkaldy on the Scottish coast, was surprised to see the open field behind her building filled with military gliders. Sentries kept her from approaching too closely, but she could tell even from a distance that there was something odd about those particular gliders. Later, studying them from an upstairs window, she realized that the aircraft were made entirely of burlap and plywood.

Farther down the winding road, a farmer accidentally wandered into a military restricted area. As he started to turn back, he noticed troops unloading bulky piles of rubber at intervals along the hedge line. A small tractor moved noisily down the row, and before his startled eyes one of the shapeless heaps quickly expanded into a full-sized army tank. Soon there were many other "tanks" ranged around the field, all made of rubber that had been inflated by compressed air.

The plywood gliders would never fly, the pneumatic tanks would never rumble into battle, but they were combat weapons just the same—opening chords in the subtle Fortitude orchestration.

At the very moment that the camouflage squads were

preparing these decoys, Hitler was meeting at Berchtesgaden with von Rundstedt, Field Marshal Hugo Sperrle, and Admiral Theodor Krancke. Worried about the problems ahead and the risks involved, he told them, "The destruction of the enemy's landing attempt means more than a purely local decision on the western front. It is the sole decisive factor in the whole conduct of the war, and hence in its final result."[1] The Nazis had no illusions; an Allied invasion was inevitable, but exactly where would the attack come? It could, the führer pointed out at that same meeting, begin anywhere, ". . . on any sector of the long western front from Norway to the Bay of Biscay, or on the Mediterranean—either the south coast of France, the Italian coast, or the Balkans."[2]

Hitler, brooding over his maps and intelligence reports, was determined to defend every inch of the German-occupied territory; but to do this he had to keep his forces widely scattered, while holding back strong reserves for last-minute contingencies. For months the führer's thinking had seesawed as to when and where the Allies would strike, and these uncertainties became a handy fulcrum on which the Fortitude planners could balance their hoax.

To exploit the enemy's dilemma—and to further sharpen their plan—the Allies divided Fortitude into two sections: Fortitude North, or "Skye," and Fortitude South, or "Quicksilver."

Fortitude North played on Hitler's obsession with the tactical and strategic importance of Norway. By the time the Allies were ready to move across the Channel, the German dictator had assembled a force in Norway of some four hundred thousand men, complete with planes, tanks, and heavy artillery. The mission of Fortitude North was to *keep* them there, out of action, and the only way to do that was to convince Hitler of what he already half believed: that a strong Allied army was massing in Scotland, preparing to attack Stavanger and then push on to Oslo.

Of course such a scheme would have been a lot easier if the Allies could maneuver with genuine divisions and armored battalions, but at that point every last man, tank, and heavy gun was being funneled into the legitimate invasion needs of Operation Neptune. So the conjurors of Skye had to improvise, creating fake divisions and simulating visual and aural realities where none, in fact, existed. Above all, the Allies had to make their frauds believable. The Kirkaldy tank replicas were realistic even at close range. They had been produced in British and American factories from templates and mechanical drawings of the originals, and were remarkably similar to M-4 Shermans and Crusaders in size, shape, and detail. Shadows are an important part of the language of photo interpretations, and the facsimile tanks cast accurate shadows when seen or photographed from the air.

The camoufleurs also used specially designed two-wheel trailers which they towed back and forth behind trucks, leaving realistic "tank tracks" in their wake. When an inflated Sherman was placed at the end of these tracks, the effect from overhead was even more convincing. As a final necessary touch, the bogus machines were draped with tree branches and camouflage netting—but this "protective cover" was deliberately left careless.

The Scottish crews had only a limited number of replicas, but working at night they periodically disassembled their gliders and deflated their tanks and set them up in new locations, to create an effect of fresh airborne and armored battalions moving into the area.

Farther to the east, in the coves and estuaries of the Firth of Forth, another element of General Thorne's "assault force" took shape. Moored along makeshift docks were masses of decoy troop barges and other landing craft, all made of canvas, wood, and chicken wire. Camoufleurs enjoy adding extra touches of realistic detail. On some barges, funnels puffed lazy columns of smoke; sailors' laundry hung on others, swaying gently in the offshore breezes; pop tunes and the BBC News blared from radios in a few of the ships' cabins; and as the sounds of "Don't Sit Under the Apple Tree" and "The White Cliffs of Dover" drifted over the shoreline, a number of real tugs and small vessels chugged energetically around the moorings of the strange chicken wire navy.

In addition to these fictions, dozens of wooden twin-engine "bombers" began to appear on airfields near Edinburgh and Glasgow. Dummy antiaircraft guns made of painted poles—with a few genuine guns interspersed—dotted the landscape. Empty containers simulating crates and drums of "equipment and fuel" grew in neat stacks at various collection points. Jeeps, trucks, and command cars raced up and down the dusty roads, following signposts directing them to "Fourth Army bivouacs." All this activity, carried out meticulously by a handful of men, gave the impression of a large military body assembling for a major offensive—an impression not wasted on the Luftwaffe.

By that spring the Allies had won decisive air superiority, but occasional Condors and Messerschmitts still made bombing or strafing raids; and on clear days a brave Luftwaffe pilot or two would fly over on reconnaissance with cameras set to record the activity below. In the Edinburgh sector, German recon planes were allowed to photograph exactly what the Allies wanted them to find. MI-6 ordered the antiaircraft crews to fire heavily at the intruders to keep them high; the somewhat puzzled men were also told to aim poorly. When a Luftwaffe recon pilot got back to his base, thankful to have made the trip safely, the aerial photos he brought back were examined with great care. Von Roenne's analysts studied the brigades of tanks, the squadrons of gliders, the trails of dust along the roads, the wakes of small ships moving buglike among the barges at Dunbar and North Berwick, Anstruther, Leven, and Buckhaven. They frowned through their magnifying glasses, made their measurements, wrote their reports—and the German intelligence dossiers soon began to grow.

But visual deception was only part of the Skye plan: "aural camouflage" was needed as well. Huge modern armies, spread over miles of terrain, depend heavily on good systems of communication. As real troops and tanks arrive in staging areas, aircraft assemble at landing strips, and supply and fuel depots are organized, the airwaves in the vicinity crackle with reports, questions, instructions, bulletins, complaints, clarifications, orders and counterorders; and this busy crisscross between various units forms a distinctive radio traffic pattern. The *Funkabwehr,* Germany's wireless and cryptanalytic service, had become adept at locating these patterns, interpreting their meaning, and pinpointing the wireless stations involved; which meant that the Allies would have to create aural signals of the greatest possible fidelity.

Because of his impressive credentials, familiar to the Germans, General Thorne had been chosen as nominal "commander" of this nonexistent army; but the actual creation of the illusions was supervised by Colonel R. M. MacLeod, a fifty-year-old veteran British officer who, the LCS had fortunately discovered, was an authority on historic military ruses. Under his expert direction a small crew of Signal Corps operators simulated an entire army communication pattern complete in every detail. Some of their messages were sent in the clear or by radio telephone while others went out in cipher, to be labored over by the *Funkabwehr's* code experts.

MacLeod's radio deception was inspired: Besides the usual day-to-day orders and instructions, the communiqués dealt with the ski training of personnel, advice on the use of snow equipment, requisitions for crampons and heavy boots, data on the performance of tank engines in below-zero weather—all the minutiae that characterize an invasion force possibly bound for Norway.[3] At Zossen these wireless messages were carefully recorded and evaluated, and von Roenne's dossiers on "BFA-Edinburgh" continued to expand.

Local Scottish newspapers also contributed to the deception. The citizens of Dundee, Perth, and Falkirk, browsing through their dailies, often came upon news items (planted by MI-6) about Fourth Army dances and pipe band concerts, battalion soccer and football games, the arrival of SHAEF dignitaries on tours of inspection, even the marriage of an officer at Edinburgh headquarters to a member of the "Seventh Corps Women's Auxiliary."

The Nazi intelligence agencies and the generals at *Oberkommando der Wehrmacht* followed all these activities with wary interest. They had been tricked before by decoys and wireless hoaxes (in fact considerably more than they realized), and even though the Scottish situation seemed legitimate, they wanted corroboration. So they turned to two trusted undercover agents working for them in that area. These agents had been flown by seaplane to the north coast of Scotland in April 1941, and had waded ashore at night. Their mission was to sabotage local food and supply dumps, and to pass along any military data that came their way. The spies, given the unlikely British code names of "Mutt" and "Jeff," had proved their worth many times over and were highly regarded by their German superiors; so it was logical for FHW to rely on them at this crucial time.

What the Germans didn't know, however, was that the two agents had surrendered as soon as they stepped ashore, and since then had been under the control of the Allies. It would have come as a further terrible shock to Colonel von Roenne to learn that not only Mutt and Jeff, but *every German agent in Britain*, was either dead, jailed, or working under tight British-American supervision.

This remarkable coup was the result of well-synchronized Allied intelligence planning. Counterespionage during the war was the joint responsibility of Britain's MI-5 and the Special Operations Branch, or X-2, of the OSS. Working closely with these agencies were the FBI in Washington, Russia's NKVD, and other units such as SHAEF's Special Means ("Ops B"), British Security Coordination, and the Pentagon's JSC, or Joint Security Control. But the specific job of converting or "doubling" enemy agents in Britain went to a unit of MI-5 known as the XX Committee—the use of the double-X in its title symbolizing the "double-cross system" which it so successfully put into practice.

In the course of the war the mysterious and little-known XX Committee, directed by Lieutenant Colonel Thomas Robertson, managed to track down and arrest scores of operatives working for German intelligence. A number of agents whose activities constituted treason were executed, others were imprisoned, and some, such as Mutt and Jeff, defected to the Allies voluntarily. Many others agreed, through threat or persuasion, to switch allegiance and cooperate with the Allies—working under code designations such as "Meteor," "Peppermint," "Treasure," "Sniper," "Garbo," "Tate," "Bronx," "Artist," "Brutus," and "Tricycle."

In his memoirs John C. Masterman, an Oxford don and member of the XX Committee, who later became its historian, explained the function of the so-called double-cross system. The primary object of any wartime security service, he pointed out, was to deny *all* information to the enemy—but the ideal could never be fully achieved since it was patently impossible to keep so many preparations and activities a total secret. "On the other hand," Masterman wrote, "you can control one of the enemy's main sources of information, and thus know what his information is—and, to go a step further, you can select his information for him, you can pervert his information, you can misinform him and, eventually, actively deceive him as to your intentions."[4]

The confidential history of the OSS, recently declassified, documented America's heavy involvement in this counterespionage work and amplified the process:

> One of the principal uses of double agents was to feed the enemy such seemingly good information from a given area that he would feel no need of sending additional agents to the region. In this fashion, X-2 could gain complete control of the intelligence which the enemy received from a particular area.[5]

Doubling an agent was a sensitive procedure calling for a careful blend of imagination, ingenuity, and good judgment. A captured spy facing long years of imprisonment was understandably vulnerable, and subtle appeals to self-interest and self-preservation could be applied, as well as less subtle but equally attractive offers of financial reward. This blend of pressures succeeded so well that Masterman was able to report, "For the greater part of the war we did much more than practise a large-scale deception through double agents; by means of the double agent system *we actively ran and controlled the German espionage system in this country.*"[6]

The XX Committee's achievement was impressive yet feasible. Great Britain was a relatively small island where tight security measures could be imposed and rigidly maintained. A secret agent had three basic ways of transmitting information: by wireless telegraphy, by written notes in code or secret ink, and through personal contacts in neutral countries such as Sweden, Portugal, and Switzerland. With a diligent, well-trained network of operatives, these methods could be monitored. In addition, every double agent had an experienced "case officer" who acted as his or her controller, confessor, and general supervisor. The case officer's job was to make sure that the agent sent those messages—and only those—the Allies wanted sent. Of course, to maintain an agent's credibility with the Germans, genuine bits of information also had to be passed along and transmitted; but most of the data flowing to FHW, *Amt Mil* and the SD were carefully tailored to fit the Fortitude scenario.

Norman Holmes Pearson, a former liaison officer with the OSS, explained how it was done:

> The techniques of intercepting messages sent by wireless were highly developed. So was the science of direction-finding by which the location of the transmitting instrument could be determined. By the constant monitoring of all wireless traffic . . . it was possible to check not only

what the double agent sent out . . . but also messages sent to him.[7]

Questions sent *by the Germans* to their agents were a great source of added intelligence. By studying these queries and messages, the Allies could learn about enemy plans and intentions and also gain information about the existence of new spies.

The biggest problems of the double-cross system were not technical but psychological; a successful XX Committee case officer, Masterman explained, ". . . had to identify himself with his case; he had to see with the eyes and hear with the ears of his agent; he had to suffer himself the nervous prostration that might follow an unusually dangerous piece of espionage; he had to rejoice with his whole heart at the praise bestowed by the Germans for a successful stroke."[8] Between agent and case officer there developed an almost symbiotic relationship; and each spy's psyche was read like a fever chart, every mood and reaction weighed in a process Pearson called the "ecology of double agency."

In the case of Skye, faked messages from Mutt and Jeff to FHW helped greatly to strengthen the overall deception. Mutt, docile and obedient, had been living quietly as a "farm manager" in Scotland. He dutifully answered the queries from his German contact, sending information on his wireless set, which had been prepared for him by his case officer. After a period of time, Jeff had become uncooperative and was now locked in Dartmoor prison; but FHW assumed he was at liberty, and his message style and transmission technique were skillfully imitated by the XX Committee.

Between them, the two agents supplied the "confirmation" sought by the Germans. Yes, the Fourth Army was definitely assembling for a Norwegian campaign, probably to come in July. Yes, it was a large force consisting of two corps, four infantry divisions, an airborne and armored division, and an armored brigade—and hundreds of ships and landing barges were being prepared for them, as aerial reconnaissance would show. The agents also reported the arrival in Edinburgh of Budyenny, the Russian liaison officer—another strand in General Morgan's fabric of "unbroken plausibility." Russia, aware of the importance of Operation Bodyguard, had been cooperating effectively. From NKVD plants and other agents, the Germans heard (and believed) that the Red Army was concentrating troops and ships at Murmansk and Petsamo, ready to drive into Norway from the north in coordination with the Allied landings.

In the gradual building of the hoax, no supportive detail was neglected. The U.S. Air Force and the RAF made extra photographic flights over the Norwegian coastline. The Royal Navy sent flotillas from Scapa Flow to "reconnoiter" the Skagerrak area. Russian submarines appeared with increasing regularity off the northern fjords. And SOE, MI-5's Special Operations Executive in charge of resistance activities, stepped up its coded messages to the Norwegian underground.

In the United States, another dimension was created by a double agent known to the FBI as "Albert van Loop."[9] A mild-mannered, middle-aged Dutch national who had been a German spy during World War I, van Loop—whose real name was Walter Koehler—was again recruited by the Nazis at the start of World War II. Arriving in America via Madrid and Lisbon in 1942, Koehler promptly defected and was placed under close FBI control in New York City. Here he continued to act out his role as a covert agent, but was actually used to channel doctored information to German intelligence. A secret radio station was set up in a remote spot on Long Island, and through this van Loop notified his contact in Hamburg that he had taken a job as "night manager" in a New York midtown hotel. The hotel ostensibly catered to American Army officers en route to England, which placed the spy in a position to obtain valuable data.

For months, van Loop's messages to Hamburg (carefully prepared by the FBI) contained valid but very trivial facts about American troop movements. Then gradually, these weekly reports focused on Iceland. The communiqués indicated that strong American forces were being assembled in Iceland for a major amphibious assault on northwest Norway, in support of General Thorne's attack farther to the south. To strengthen this deception, fake troop barracks were built in the Reykjavik area and dummy barges appeared in Icelandic waterways. The Germans considered Koehler an extremely useful agent, and great attention was paid to his dispatches.

Years later Ladislas Farago, the historian and authority on espionage, found evidence indicating that the devious Koehler was actually a *triple* agent who had managed to outfox the FBI, and to send a great many uncontrolled messages to a secret German contact in occupied Paris. Whether this was actually the case, and whether in Farago's words Koehler was "the spy who fooled J. Edgar Hoover," the fact remains that his "Iceland reports," sent under FBI supervision, helped to bolster the Skye invasion hoax.

Even neutral Sweden played an unwitting part, as American and British agents circulated through Göteborg, Stockholm, and Malmö, discreetly (but not too discreetly) collecting data on Swedish rail lines and port facilities, and initiating rumors about the coming Allied operation. Whispering campaigns prompted the Swedish people to stock up on food, wood, and fuel oil, in case the Norway attack escalated and led to air raids on Swedish supply and transport centers. And as a final touch, the British Treasury—at Colonel Bevan's request—began to invest heavily in Scandinavian securities, heightening the impression that the Allies soon expected to return.

Any of these deceptions, taken singly, might have seemed thin or inconclusive to the Germans, but as the structure grew its many elements confirmed and reinforced each other. Allied security was so effective that, even with an occasional misstep, the desired *Feindbild* emerged; and the anxious von Roenne, shuffling his aerial photos and secret reports, was able to advise OKW with conviction that the British Fourth Army in Scotland constituted a serious threat. Hitler, receptive to any opinion that bolstered his sense of intuitive genius, reacted with satisfaction. He had always considered Norway his "zone of destiny," and now, he told his generals, the Wehrmacht divisions would remain on duty there regardless of anything that happened elsewhere.

The decoy tanks and gliders, the dummy landing barges, the fake wireless traffic, the doctored reports from Mutt, Jeff, and the other agents, had done their work. The German dictator was so convinced of the rightness of his Norway decision that he stubbornly refused to change it, even when Allied armies—supported by tanks made of heavy steel instead of harmless rubber—began driving inland from the conquered beaches of Normandy.

But long before the Neptune attack, while the sham divisions of Thorne's army were just beginning to "mobilize" near Edinburgh, another drama was unfolding miles away—a singular piece of deceptive theater in which, appropriately, the Shepperton Film Studios played a part.

Shepperton, one of England's leading motion picture companies, has produced many notable films, and its talented scenic designers have created everything from Mycenaean temples and three-masted schooners to elegant Edwardian salons and the narrow streets of Covent Garden. But a few months before D-day they took on an assignment more difficult than they ever faced on a movie set. At the request of Bevan and the LCS, the designers built a giant oil storage facility and docking area near Dover on the English Channel. This installation, stretching along several miles of shoreline, consisted of pumping stations, pipelines, storage tanks, jetties, truck bays, terminal control points, troop barracks, and antiaircraft defenses.

The new oil docks were greeted with a show of official approval; King George VI and General Bernard Montgomery both paid inspection visits, which were noted in the press; and General Eisenhower attended a special celebration dinner for the project's "engineers and construction workers," held at the White Cliffs Hotel in Dover. The whole facility, carefully guarded by military police, was of course a complete fraud. The set designers and camouflage crews had built it entirely of painted canvas, wooden scaffolding, fiberboard, and sections of old sewer pipe—and the only real oil in the vicinity rested in the crankcases of army trucks used to haul in the necessary construction materials.

An invading force obviously needs vast amounts of fuel, and the Germans were led to believe that the elaborate Dover complex was one terminus of an underwater oil pipeline that would eventually extend to Calais in France. This deception was part of the Fortitude South (Quicksilver) scenario, activated at roughly the same time as its Scottish counterpart. The two plans had similarities, but Fortitude South was far more urgent than Fortitude North, because it dealt with *immediate geographic dangers to the real invasion*.

For months the Germans had shifted their divisions around Europe like so many lethal chessmen, trying to anticipate when and where the Allies might strike. An assault on Norway, they believed, was highly possible, but the threat to occupied France was even more serious. Hitler and most of his generals expected a major attack to come somewhere on the Pas de Calais, an area stretching from Dunkirk west to Calais, rounding Cape Gris-Nez and curving south to St. Valery at the mouth of the Somme River. Von Rundstedt, commander in chief of the Wehrmacht's western forces, tapped the big map on the wall of his underground bunker at Saint-Germain-en-Laye and announced to his staff officers, "If I were Montgomery, that is exactly where *I* would attack."[10]

The *Feldmarschall*'s reasons were logical. The shortest passage across the choppy waters of the Channel lay in a twenty-one-mile line between Dover and Calais. A journey from the south of England to the Normandy shore would take three or four times as long, and one to Brittany would take even longer. The Calais coast was also the quickest route to the Ruhr valley, where the core of Germany's heavy industry was located. Von Rundstedt knew this would be a high priority target for the Allies, but if they landed in Normandy or Brittany they would have to fight their way across half of France before reaching the Ruhr.

An invading army also needs adequate ports in order to land tanks, trucks, artillery, ammunition, food, medical supplies, and other equipment. There were no useful harbors on the Normandy coast except for Le Havre and Cherbourg, both of which were very heavily fortified and far from any viable landing beaches. But a bridgehead on the Pas de Calais would put the Allies close to the ports of Boulogne, Dunkirk, and Calais itself; and not too distant was Belgium's great port of Antwerp.

There were other reasons for von Rundstedt's assumptions. Allied fighter planes, limited in range, could protect an invasion much more effectively if landings were made close to their air bases. Furthermore, the Calais area contained the secret V-1 missile launching sites being readied for massive attacks on Britain. It was logical to assume that the British and Americans would try to capture these sites before German rockets destroyed London and other coastal cities. The Allies knew all about the missile sites and had bombed them repeatedly; in the winter of 1943, most of the launching platforms had been put out of action, but some had been repaired and many new ones, well camouflaged, were being secretly added. Gilles Perrault, historian of the French resistance movement, summed up the Wehrmacht's view. They were certain, he wrote, that the Allies ". . . would not let such a deadly menace hang over London; surely they would attack the area where Germany's reprisal weapons were concentrated."[11]

If the Allies *did* attack the Pas de Calais, as expected, the Nazis would be ready. By the spring of 1944, Hitler had shifted an estimated sixty divisions to western France. They had been spread thinly over wide areas, but the greatest concentrations were in and near Calais. Stationed here was the powerful Fifteenth Army, which included two large armored divisions—the Panzer Lehr, located at Le Mans, considered the strongest in the German army, and the fanatical Twelfth SS Hitler Jugend, centered at Lisieux. The laborers of the "Todt Organization," under Field Marshal Rommel, had built one hundred thirty-two heavy concrete gun positions between Dunkirk and Boulogne, compared to forty-seven along the Normandy beach line. And echeloned behind the Calais front were many strong reserves—far more than on the Cotentin or Breton peninsulas.

"If the Atlantic Wall anywhere resembled the ferocious image popularized by Goebbels," Perrault noted, "it was in the segment where von Rundstedt expected the Allies."[12]

But the Allies had no intention of obliging him. The Combined Chiefs of Staff had long since decided on Normandy as their invasion target; and had solved the difficult supply problem by constructing two huge movable ports called "Mulberries," which consisted of enormous concrete floating caissons joined together. At the proper time, these would be towed to Normandy and anchored off the beaches to serve the liberating armies as artificial harbors. The true purpose of the Mulberries had been successfully kept from the Nazis, which raised Allied hopes of gaining a degree of surprise on D-day.

Now the goal of Fortitude South was to protect those fragile hopes by exploiting German fears about the Pas de Calais: Everything would have to be done to convince them that they had guessed correctly—and even after the landings were made, they would have to believe that Normandy was only a diversion and that a bigger blow would fall between Dunkirk and the Somme. Only in this way could von Rundstedt's heavy armor be kept far enough from the target beaches. As Masterman explained:

> By the early spring of 1944 it was utterly impossible to disguise the fact that the major attack would come

> somewhere between the Cherbourg peninsula and Dunkirk; the true preparations, which could not be wholly disguised, indicated this beyond all doubt, and the distance from the base at which fighter cover could be supplied helped define the limits. The deception policy was dictated by these circumstances. . . .[13]

So the Quicksilver plan had, in fact, several goals: to conceal the real date of the invasion, to indicate a false invasion area, and finally to convince the enemy (after the blow fell) that another and greater attack would come elsewhere.

The Dover "pipeline," part of this intricate hoax, was designed by one of England's leading architects, Professor Basil Spence, whose sketches and blueprints were followed by the set-builders of Shepperton. Ground and air defenses kept the Luftwaffe away during construction of the docks, and afterward the results were tested. "Most of us were film and theater people," one of the camoufleurs recalled, "so naturally we wanted a proper dress rehearsal." The RAF sent a low-level photo plane over the Dover stage set, the aerial prints were then examined, and necessary changes were quickly made. And at that point the defensive screen was gradually relaxed.

Spence's counterfeit docks proved to be good box office. German planes came over periodically to photograph them; but fighter patrols and antiaircraft kept the intruders at altitudes of thirty thousand feet, and at that height it was virtually impossible for enemy cameras to pick up any remaining flaws. On German prints, the Dover docks looked authentic. Now and then, Nazi long-range artillery on Cape Gris-Nez would even lob a few inaccurate shells at the terminus—and whenever these landed the camouflage crews would create suitable "fire damage" using sodium flares and mobile smoke generators.

Quicksilver, directed by Colonels J. V. B. Jervis-Reid and Roger Hesketh, expanded rapidly, and throughout East Anglia—as in the Edinburgh area—a huge phantom army soon came into being. It was known as the "First United States Army Group," or "FUSAG," and it had on paper its own command staff, order of battle, headquarters corps, recognition signals, and a full complement of troops, aircraft, and armor. FUSAG theoretically consisted of the U.S. Third Army and the Canadian First Army, backed by scores of divisions still training in the United States. According to the Quicksilver script, twelve divisions were to take part in the "Calais landings" and an additional thirty-eight would be sent in after the beachhead was secured. Ostensibly, the U.S. Ninth Air Force would provide tactical support for the invading troops.

Many of these FUSAG divisions were, of course, fictitious, but others were quite genuine and scheduled for action in Operation Neptune. This careful blend of real and notional troops served a threefold purpose: it cloaked the D-day buildup, disguised the Allies' real destination, and at the same time strengthened the credibility of the hoax.

And now, on the center of the Fortitude stage stepped a military leader well accustomed to the spotlight—FUSAG's figurehead commander, General George S. Patton, Jr., whom Ladislas Farago characterized as "the tall, taut, tense American general of vague fame and growing notoriety, the swashbuckling tank wizard the tabloids in the States already called 'Old Blood and Guts.'"[14]

Hitler referred to him contemptuously as "the cowboy general," but Patton was greatly respected—even feared—by the German High Command. Von Roenne and the other intelligence officers were impressed with the dynamic leader's military capabilities, and felt certain that he would have an important role in the invasion. Actually, Patton had returned to England under a cloud because of incidents in Sicily involving the slapping of two hospitalized soldiers. This had caused a great furor but Eisenhower, rightly considering Patton indispensible to Overlord, had assigned him command of the U.S. Third Army, a force still training in the United States and not slated to join the European campaign until well after the initial landings. In the meantime, while waiting for his troops, "Old Blood and Guts" would serve the Allies as their most distinguished decoy.

According to Farago, Patton's biographer, the general was deeply disappointed at being excluded from the invasion plan, but he recognized the importance of Quicksilver and played his part with flair and competence. Patton's FUSAG "appointment" neatly finessed the overall deception, and soon FHW and *Amt Mil* were referring in their briefing papers to *Armeegruppe Patton* which, they reported, was assembling for an assault across the Straits of Dover. Patton and FUSAG later paid extra dividends. When the Allies finally did invade Normandy, the Germans learned that Patton wasn't among the Allied commanders in France. This strengthened their belief that Normandy was a diversion, and that Patton had been saved to lead the main invasion later on.

George Patton's "command," like Andrew Thorne's, was backed by a great show of military activity. The Shepperton designers again went to work, creating in addition to their oil docks hundreds of landing barges made of canvas and wood that floated on empty oil drums. The fake vessels, resembling Allied LCTs, were known somewhat coyly as "bigbobs." Other invasion craft of inflatable rubber, called "wetbobs," were also produced. Some of these dummy barges were moored in creeks and inlets between Great Yarmouth and the Thames estuary, but the majority were concentrated in the Ramsgate-Hastings area opposite the Pas de Calais. Soldiers from over-age battalions served as "crews" for these vessels, and authentic details were supplied. Smoke drifted from the barges' stacks, naval tugs towed them busily to and fro, and around some of the ships patches of "spilled oil" appeared, which would show up well on German aerial photographs. At night the camouflage squads used Q lighting to represent military docks; they also ran carts with shielded "truck headlights" along the shore roads, simulating fueling and loading activity.

As in Scotland, dozens of inflated rubber tanks were moved into coastal "staging areas" to prepare for the campaign. At that point, according to reports, a meadow near the little town of Chaul End became the setting for a *corrida* unlike anything ever seen in Barcelona or Madrid.[15] A large bull accidentally got loose in a field where a platoon of rubber tanks had been stationed. The puzzled animal charged one of them, sank in a sharp horn, and the punctured tank ingloriously collapsed—the only case on record in which a Fortitude decoy took part in actual combat.

To give *Armeegruppe Patton* full dimensions, radio camouflage was again used. FUSAG wireless traffic buzzed endlessly between Wentworth and Chelmsford, Leatherhead and Dover, Folkstone and Bury St. Edmunds, duplicating the busy network of an army in mobilization. Sudden periods of "radio silence" were also imposed, to intensify the war of nerves and further puzzle the Germans. Security was heavily stressed; in March 1944, an Allied Headquarters' directive to FUSAG's radio personnel stated:

> You are taking part in an operation . . . designed to deceive the enemy, and it has a direct bearing on the success of our operations as a whole. You must realize that the enemy is probably listening to every message you pass on the air and is well aware that there is a possibility that he is being bluffed. It is therefore vitally important that your security is perfect; one careless mistake may disclose the whole plan.[16]

These fake communiqués were all designed, of course, to dangle clues in front of the *Funkabwehr*. One such message became a SHAEF classic, still remembered and fondly quoted: *Fifth Queen's Royal Regiment report a number of civilian women, presumably unauthorized, in the baggage train. What are we going to do with them—take them to Calais?*[17]

Mutt and Jeff, the double agents in Scotland, also had their counterparts in the drama. To reinforce the physical deceptions, the XX Committee used many operatives to feed information to the Germans; but the bulk of this work was accomplished by three agents known as Brutus, Garbo, and Treasure.

Brutus, a member of the Polish underground, had been captured by the Nazis in 1941 and (so they believed) was eventually pressured into becoming a German collaborator. They arranged for him to "escape" and make his way to England, where he promptly contacted the XX Committee. Garbo, a Spanish national who lived in Lisbon, was supposedly an importer of fruits and vegetables, and his business involved contact and trade between England and Spain. A tireless, energetic spy, Garbo ostensibly had a "private network of agents" in England supplying him with information, which he dutifully passed on to FHW. The information came, of course, straight from X-2 and MI-5. Treasure was a young Frenchwoman whose uncle had been a general in the Imperial Russian Army. Because of her background and supposed "Czarist leanings," she gained the confidence of the Germans; once settled in England, she let them know that she was having a romance with a FUSAG officer, which put her in a position to pass along data of value.

Other double agents who played key roles at that time included Tricycle, a Yugoslavian businessman who traveled with military permission between London and Lisbon, and Bronx, a South American woman whose father served with the Argentinian Embassy in Vichy. Bronx had supposedly managed to join the British Women's Army Auxiliary, from which vantage point she could obtain and transmit useful information.

All these agents, closely supervised by their case officers, supplied the Germans with facts—some true and many false—which bolstered the illusion. They informed von Roenne, among other things, that the Allies had set D-day for the end of June or early July—weeks later than the actual calendar date. They also supplied the Germans with a FUSAG "order of battle," giving the exact makeup and geographic distribution of Patton's chimerical troops. To the methodical Germans, any list of hard facts and figures would seem highly persuasive. Quicksilver's faked order of battle was circulated widely by the Wehrmacht's intelligence agencies, and became a fixed reality in the thinking of the German commanders. The end result, Masterman reports, was satisfactory:

> A German map . . . as of 15 May 1944, which was later captured in Italy, showed how completely our imaginary order of battle had been accepted, and was largely based on information supplied by the double-cross agents, especially Garbo and Brutus.[18]

The work of Garbo in particular was so greatly appreciated by the Germans that they decided in June 1944 to award him the Reich's Iron Cross II *in absentia* for his valuable services to the führer. Later that year the British awarded him the M.B.E. (Member of the Order of the British Empire), because of his role in perpetuating the hoax, giving Garbo the peculiar distinction of having been decorated by both sides of the same war.

Other military elements also played their part in the Quicksilver masquerade. Allied destroyers, PT boats, and minesweepers maneuvered purposefully off the East Anglian coast and sortied out into the Channel. Naval and air units used special devices to jam Nazi radar. Sabotage by the *maquis* and other resistance groups increased everywhere, but especially in the Calais area, where German patrols were ambushed, troop trains derailed, radio stations attacked, telephone lines cut, supply dumps destroyed, fuel depots set on fire. Even the Allied bombing program, vital to Neptune, was adjusted to Fortitude. In preparation for D-day, the U.S. Eighth and Ninth Air Forces and the RAF were responsible for interdicting all bridges and rail lines in Normandy in order to wreck the Wehrmacht's supply network. But to effectively disguise this, a "two-for-one" plan was adopted: for every bombing mission sent out over Normandy, *two* equivalent missions were scheduled for the Pas de Calais.

This two-for-one plan, a classic Houdini misdirection, was a distinct success. "Allied air activity," Cave Brown wrote, "had increased greatly, and the Germans detected that its main thrust was the destruction of supply lines. But the campaign was carried out in such a way that they could not deduce where the Allies would land."[19] And a mere forty-eight hours before D-day, with the Normandy-bound convoys *already at sea,* Rommel in his weekly report to von Rundstedt stated, "Concentrated air attacks on the coastal defenses between Dunkirk and Dieppe [i.e., the Pas de Calais] strengthen prospects of a large-scale landing in that area."[20]

The deception plan received further help from a highly improbable source—the Wehrmacht itself. During the early war years, German military intelligence had been the responsibility of the *Abwehr,* a huge agency directed by a complex, shrewd, and enigmatic intellectual, Rear Admiral Wilhelm Canaris. The mysterious Canaris, once highly esteemed by Hitler and OKW, gradually fell into disfavor, suspected—with valid reason—of "defeatism" and anti-Nazi sentiments. A powerful and aristocratic naval officer of the old school, Canaris secretly detested Hitler and had done much to thwart and undermine the grotesque Nazi leadership; but his intrigues grew too obvious, and in February 1944 the *Abwehr* was abruptly disbanded and its director banished to an obscure post in the Reich government.

The German intelligence service was in chaos, and a new structure was formed under von Roenne of FHW and Schellenburg of *Amt Mil,* both reporting to Ernst Kaltenbrunner's SD, the security branch of Hitler's elite personal army. There was hostility and distrust between these various groups, and the sudden fall of Canaris led to great bureaucratic confusion and inefficiency—all of which played nicely into the hands of the Fortitude manipulators. The Germans, already staggering under mountains of puzzling claims and counterclaims by double agents, plagued by brilliant camouflage decoys and superb wireless fakery, groped through an endless maze of illusions, dead ends, feints, and false trails; and inevitably von Roenne and his associates, unable to filter out the sham from the valid (which would have been difficult under the best of circumstances), slipped deeper and deeper into the trap which had been prepared for them.

When this trap was finally ready to spring, even the weather conspired to help the deception. For the huge amphibious operation to succeed, weather conditions had to be just right; but at the beginning of June the English Channel, always somewhat unpredictable, turned stormy and turbulent. The D-day assault had originally been set for dawn on June 5, but as the time neared reports from SHAEF's Meteorological Committee were discouraging. Eisenhower recalled:

> Low clouds, high winds and formidable wave action were predicted to make landing a most hazardous affair. The meteorologists said that air support would be impossible, naval gunfire would be inefficient and even the handling of small boats would be rendered difficult.[21]

The risk, Eisenhower decided, was much too great; major units already under way were secretly recalled and the vast undertaking was postponed for twenty-four hours. However, weather conditions continued to grow worse and so did the gloom and tension at Allied Headquarters. Then, quite unexpectedly, the forecasters found surprising indications of a short letup in the bad gales. The respite, they predicted, wouldn't last very long, but there *would* be enough time for Neptune to be launched successfully. With this slim margin of meteorological safety to lean on, Eisenhower and his commanders gave the final signal to go ahead.

The Germans, of course, had also been monitoring the Channel weather; the negative reports had lulled them into a sense of security and they were unaware of the sudden expected change. "The entire German high command," Brown reported, "was unanimous in its judgment on the eve of D-day: whatever the Allies intended, the weather was too bad to permit an invasion during the first week of June."[22] Rommel was so convinced of this that at daybreak of June 5, he left the front and headed toward Germany by car for a brief holiday with his family at Herrlingen. Other key Wehrmacht officers went off to attend war games at Rennes, in central Brittany; and at that point, General Walter Warlimont of German Supreme Headquarters noted in his diary, nobody at OKW had the slightest idea that ". . . the decisive event of the war was upon them."[23]

At 6:30 A.M. on June 6, 1944, hundreds of thousands of Allied troops, supported by a vast air and naval armada, began pouring ashore on five Normandy beaches: "Utah," "Omaha," "Gold," "Juno" and "Sword." It was an assault Churchill called "the most difficult and complicated operation that has ever taken place."[24] To increase the enemy's confusion, Allied ships again used devices to jam their radar, and hundreds of dummy "paratroopers" were dropped behind the lines near Marigny, Fauville, and Caen.

The defenders, as had been hoped, were surprised and caught off balance. Despite tough, stubborn resistance by the German Seventh Army under Field Marshal von Kluge the Allied troops fought fiercely to claw their way ashore, and before long a strong foothold had been established along a thirty-mile front.

And now, at this most critical of all points, it became evident that the bait placed so carefully in the Quicksilver trap had been swallowed. For weeks after D-day, Hitler and OKW (encouraged by continuing deceptions and false reports from their double agents) remained convinced that other landings would yet take place; and the führer stubbornly refused to free the panzers of his Fifteenth Army for use in the beachhead fighting. It wasn't until *late in July* that these powerful reserves were finally committed to Normandy, and by then the liberators were firmly established in strength.

In all this, what was the genuine contribution of the phantoms of Fortitude; and how big a part did legerdemain really play in the success of the Allied landings?

Hanson Baldwin, military analyst of the *New York Times*, wrote later, "During the assault the German command was confused and misled . . . and even prior to the assault German intelligence had vastly overestimated the Allied amphibious capability and divisional strength—a tribute to the . . . pre-invasion cover and deception techniques."[25] Eisenhower, in his SHAEF report, put it more bluntly: "The German Fifteenth Army, if committed to battle in June or July, might possibly have defeated us by sheer weight of numbers."[26] The enemy view was summarized by General Bodo Zimmermann, chief of operations at the Wehrmacht's western headquarters: "When we warned [OKW] that if we didn't get the panzers the Normandy landings would succeed . . . we were simply told that *we* were in no position to judge—that the main landings were going to come to an entirely different place anyway."[27] To which can be appended Churchill's own comment, "Our deception measures both before and after D-day had aimed at creating this confused thinking."[28]

The Nazi dictator was apparently the most confused of all. Albert Speer, his close associate and the Reich's minister for war production, recalled later how Hitler ". . . had kept on saying that the enemy would probably begin with a feigned attack in order to draw our troops away from the ultimate invasion site."[29] During the following days and weeks Speer attended numerous conferences with the führer, and found him still insisting ". . . that the invasion was merely a feint whose purpose was to trick him into deploying his defensive forces wrongly."

And finally General Omar Bradley, commander of the U.S. First Army during Neptune, said in his memoirs:

> While the enemy's Seventh Army, overworked and understrength, struggled to pin us down on the beachhead . . . the German High Command declined to reinforce it with troops from the Pas de Calais. There, for seven decisive weeks, the Fifteenth Army waited for an invasion that never came, convinced beyond all reasonable doubt that Patton would lead the main Allied assault across the narrow neck of the Channel. Thus . . . the enemy immobilized nineteen divisions and played directly into our hands in the biggest single hoax of the war.[30]

Bradley's seven decisive weeks were precisely that. By July 30, a huge Allied army was entrenched in the thick hedgerows of France's *bocage* country; the force totaled 1,566,300 men, 332,000 vehicles, and over 1,600,000 tons of supplies—all of it completely real.[31]

In that as in every battle, the victory was conclusively won by the thousands of men who fought in it; yet camouflage, subterfuge, and imagination contributed their share, and never was a *ruse de guerre* applied more effectively. Seen in historic perspective, the scheme was a major gamble involving grave risks. Rommel in particular was a shrewd, perceptive tactician with a remarkable and uncanny instinct for guessing his enemy's moves. If he had pierced the Allies' cover plan it could have been fairly easy for him to arrive at the true facts about the Normandy assault; but nowhere in German military dispatches are there any indications that the Calais "invasion" was thought to be a hoax. FUSAG and *Armeegruppe Patton* were accepted by OKW as definite realities, and other operations were looked on as purely diversionary. A study of SHAEF documents shows further that the Bodyguard planners hoped at best to delay the German panzer units for ten to fourteen days after the initial landings; instead of that, the ruse helped give the liberators seven full weeks in which to consolidate their foothold in France.

The covert art of warfare-by-illusion, its origins going back to the Trojans and even earlier, reached its peak in June 1944 on the violent, littered beaches of France. For those involved, the preinvasion scheme was a masterpiece and a culmination, but other unique forms of camouflage and deception had been used long before D-day, and often with great success.

That final road to Normandy had been paved by many earlier battles and campaigns; and Fortitude, for all its brilliance, was neither the first nor the last of the Allied hoaxes that helped to outwit the Axis armies and ultimately to defeat them.

4

Q Lights, Firewood, and Old Rubber Tires

1940–1941

The letter to RAF Fighter Command Headquarters at Stanmore, dated July 22, 1940, was brisk and businesslike:

> Sir—
>
> The first batch of dummy bomb craters were not very satisfactory, and I am now having the second batch tested at Farnborough. We have not forgotten them and they will start being issued shortly . . .[1]

The Battle of Britain had two sides: one secret and one highly visible. In the hidden struggle dummy bomb craters were of surprising value; and the July communiqué, signed by Colonel Sir John F. Turner, concerned a plan that became one of the most effective of the war's deceptions.

Eight weeks before—overwhelmed by von Rundstedt's and Guderian's panzer attacks—the British Expeditionary Forces had fallen back in defeat on the Channel coast of northern France. To meet this crisis, 850 English vessels of all shapes and sizes converged on the beaches of Dunkirk and began evacuating the shattered divisions; and within a few days, while Hitler equivocated, the makeshift fleet managed to rescue the bulk of the British troops in Europe. But the disaster for England was monumental, and on June 4, Churchill reported to a worried Parliament, "We must be very careful not to assign to this deliverance the attributes of a victory. Wars are not won by evacuation."[2]

In France, the B.E.F. had left 120,000 vehicles, 2,300 heavy artillery pieces, 8,000 Bren guns, 90,000 rifles, and 7,000 tons of ammunition behind them.[3] William L. Shirer, in his history of the Third Reich, noted that the British predicament after Dunkirk,

> . . . was indeed grim and more dangerous than it had been since the Norman landings nearly a millenium before. It had no Army to defend the islands. The Air Force had been greatly weakened in France. Only the Navy remained, and the Norwegian campaign [in April of 1940] had shown how vulnerable the big fighting ships were to land based aircraft. Now the Luftwaffe bombers were based but five or ten minutes away, across the narrow Channel.[4]

The Luftwaffe was Hitler's key to "Operation Sea Lion," Germany's plan for finally invading and conquering the British Isles—a military goal unrealized since the battle of Hastings in 1066. In preparation for Sea Lion, Hermann Goering's airmen were now called on to deliver the *coup de grâce* to Britain's defenses by destroying her air power. It was the classic domino theory: Once the RAF was eliminated, the powerful Royal Navy would have no protective air cover and could be effectively neutralized. With the navy held at bay—at least temporarily—there would be time for Wehrmacht troops to cross the Channel, fall upon the weakly defended coastline, and win a bridgehead; and from there the German armor would drive inland. But first and foremost, before this assault could begin, the Luftwaffe had to win air supremacy.

The operation to crush England, leading to the historic

Battle of Britain, began early in July 1940 and reached a crescendo during the following months. Goering's primary objective was to smash the Royal Air Force through a series of major fighter actions in which the Luftwaffe would (so he believed) enjoy great numerical and tactical advantage; in addition, Germany would cripple Britain's ability to recuperate by massive bombing attacks on her flying fields and aircraft factories.

The Luftwaffe pilots, fresh from earlier victories, were confident, well-trained, and aggressive; they had gained valuable combat experience in Denmark, Norway, the Low Countries, and France and were eager to take on this new enterprise. The British, seriously short of fighter interceptors and the men to fly them, lacking maintenance equipment and antiaircraft defenses, somehow had to hold on desperately and stave off the hammer blows while the RAF gradually built its strength.

Helped by a new invention called radar, and a superb early-warning control system directed by Air Chief Marshal Sir Hugh Dowding, the RAF met the challenge vigorously. In the air the British pilots were more than a match for the Germans; but as the days and nights went by, attrition inevitably set in. The tempo of the enemy attacks increased, many valuable Hurricanes and Spitfires were lost in the great battles, and the overworked RAF fliers neared the point of total exhaustion. British airfields were also being pounded, which became a serious problem since every field knocked out of action left a crucial gap in Dowding's interceptor network.

For Great Britain, the balance between failure and success was growing very shaky. To help tip that balance, to relieve the pressure, to somehow dilute the enemy bombing attacks even if only marginally, the British fell back on a device as old as military history: the use of decoy targets. These were particularly needed during night hours when Britain's radar and other defense measures were least effective. The decoy plan—launched tentatively and without high expectations—was directed by Sir John Turner, an air staff officer of such enterprise that he almost singlehandedly created a remarkable hoax, which grew in value and impact literally from hour to hour.

Turner's first decoys—which he had used experimentally before Dunkirk—were crude installations consisting of two rows of parallel flares, simulating emergency air strips.[5] The flares, lighted manually, were set in open countryside a mile or two from airfields known to be targets of the Luftwaffe. Born out of desperation, the ruse seemed insignificant; but to everyone's surprise (except perhaps Turner's), the Germans took the bait and the fake "runways" began to attract night bombers. Like deadly moths scouting through the darkness, enemy planes were drawn toward the burning flares. Junkers, Heinkel, and Dornier pilots, dodging flak and searchlight beams, alert to the sudden arrival of Dowding's interceptors, were only too glad to zero in on the likely target, drop their bombs accurately, and head back to their bases in northwest Europe. Messages soon reached the Air Ministry from flare sites in Kent and East Anglia reporting that they had been bombed; and a decoy at North Tuddenham, protecting the field at Watton, was even hit two nights in a row.

Encouraged by this, Sir John quickly began to expand his decoy network. There were, of course, objections from RAF conservatives: Turner's "dabbling," they felt, was a needless distraction, a waste of manpower and materials, a complication that might confuse not only the enemy but RAF pilots as well; but Turner and his assistant, Group Captain A. G. Board, weathered the criticism and won approval to go ahead.

To preserve secrecy—and give Turner as free a hand as possible—his project was detached from the normal chain of command and became known simply, if mysteriously, as "Colonel Turner's Department." In the days and nights that followed, as the air battles raged overhead, "Colonel Turner's Department" was everywhere at once—irritating the purists, bolstering the weary pilots, and repeatedly tricking the Luftwaffe. And Winston Churchill, that firm believer in camouflage and legerdemain, monitored it all with great satisfaction.

The British, at this crisis in the war, were short of fighter pilots and aircraft but not of lighting equipment, and the original crude flares were soon replaced by strands of electric lights, shaded to simulate the dim outlines of operational fields. The patterns, called "Q lights," became fairly elaborate: In addition to runway markers, many of the decoys had red obstruction lamps, landing "V"s, taxi aprons, and recognition beacons; and on some, floodlights mounted on wheels were pulled along the runways to resemble the "landing lights" of incoming planes. For the sake of simplicity, the Q patterns imitated small satellite fields rather than larger bases with their complexity of hangars, machine shops, fuel storage tanks, and administrative buildings.

One RAF veteran recalled later, "On the ground in the daytime, the decoys looked pathetic. Just a collection of old wooden poles with a tangle of wires and electric bulbs strung here and there. But when we flew over the same spot at night, the effect was amazing. It was quite impossible to tell a fake from the real thing."

Turner's night lures soon numbered in the hundreds, and each one had been positioned with great care. Wherever feasible a decoy was laid out in a direct line with enemy flight approaches. Each site was located roughly 1,800 to 3,000 yards from a "parent" airfield, but in an area that wouldn't endanger nearby towns, ports, rail yards, or industrial centers. In a number of cases, civilians were evacuated from sectors where decoys might create a possible hazard. A few of the Q sites served as free-standing targets of opportunity for enemy raiders, but most were paired with specific fields. The site at Broomfield, for example, was built to protect an RAF base at Maidstone. The Dipford site covered an airfield at Taunton. Upton covered Ossington, Soham covered Waterbeach, Lenham covered Detling, Houghton was paired with a field at Middle Wallop. Large air bases at Canterbury and Chelmsford were protected by two sites each, and a major installation at Plymouth was shielded by no less than *seven* different decoys.

With this incredible profusion of inviting targets, glowing and glimmering in the night from the south coast of England to the Midlands, it isn't surprising that the confused pilots of Sperrle's and Kesselring's *Luftflotten* frequently dropped their bombs on empty fields instead of their intended objectives. But the Germans weren't the only ones who were misled.

As had been feared, some RAF pilots—particularly in the first hectic weeks—mistook the Q sites for real runways and tried to make night landings.[6] Since many of the lights had been strung across marshes and plowed fields, through ditches and hedgerows, there were a number of disastrous crack-ups. Turner's camoufleurs, believing in their program, worked hard to devise some kind of warning system. They tried signalling with red flares and Aldis lamps, blinking their lights rapidly, or shutting them down completely when friendly planes were heard overhead, yet none of these methods proved adequate. Protests and complaints began to mount, but a workable code was finally developed: It consisted of a perpendicular bar of nine red lights forming the top of a "T" and stretching across one end of the fake runway. Since the red T-bars were reasonably consistent with normal airfield operation, they failed to arouse German suspicions. The British pilots were briefed to watch for this particular code signal, and the accidents quickly came to an end.

Ironically, some of the protesters later became so impressed with Turner's huge network that they wanted the sites adapted for use as actual emergency runways, causing the patient colonel to write numerous tactful letters explaining the basic purposes and principles of decoy targets.

As the Luftwaffe campaign intensified, the Q sites were hit again and again, siphoning off hundreds of bombs that would otherwise have exploded on genuine targets. Then another authentic touch was added. In a memo to Turner,

Headquarters Coastal Command suggested that he provide "fire damage" after each German attack, ". . . in order that the enemy should not be depressed by his lack of success."[7]

Colonel Turner welcomed the idea—not out of concern for the Nazis' feelings, but as a means of improving the hoax—and his crews began experimenting with portable "fire baskets," which could be rapidly installed at the sites. These fire simulators, later code-named "Starfish," consisted of open metal frameworks filled with scrap lumber, roofing felt, creosote, and firewood, all of it triggered by small incendiary bombs set off electrically. Whenever these combustibles ran out, the crews found that old rubber tires made an excellent if odorous substitute.

Turner's Starfish were set up at various points around the decoys, and soon the *Staffel* pilots had the satisfaction of seeing splendid fires raging in the wake of their attacks.[8] The flames, leaping into the darkness over Sussex or the North Downs, were impressive and, of course, were duly reported by the German fliers when they returned to their bases.

Innovation in warfare (particularly in the area of camouflage) often leads to the need for still more innovation. The surprising success of the Q sites soon created new problems for Colonel Turner; and it was at this point that he began to think about dummy bomb craters.

The Germans, anxious to check the results of their bombing program, had begun to send daytime photo planes over the British targets; and this, Turner realized with alarm, might place the whole decoy network in jeopardy. A squadron of Me-109s, for instance, could be sent out to dive-bomb the RAF base at Feltwell. Instead, they might raid a nearby Q site at Lakenheath, and return home to report that they had done considerable damage. But photographs taken the following day, and rushed to Zossen, would show the airfield at Feltwell completely intact and still operating normally. If this happened once or twice, it could be simply attributed to human error; but if it occurred repeatedly, it would become obvious to the Germans that they were being duped. A systematic check of flight routes and target maps could help them to pinpoint these traps, and the decoys would lose their value.

John Turner, his precise military mind tempered by both humor and imagination, wrestled with this new crisis. How could he safeguard the effectiveness of the endangered Q system? His solution to the problem was a simple one: To protect the sites and go on deluding the Germans, *fake damage* had to be created at the real airfields. Convincing bomb wreckage, he knew, could be simulated in part by scattering piles of rubble and by displaying crashed and burned-out planes (many of which were unfortunately available). But more important, realistic bomb craters had to show up on the German aerial photos, pockmarking the runways, tarmacs, and taxiing aprons of the target fields.

Dummy craters were the answer, and the artists of Colonel Turner's Department were soon mass-producing them. The "craters," painted on large sheets of canvas, came in two styles: one for cloudy days and one for sunny weather. The "cloudy" model had subdued shadows. The "sunny" version had deep, sharp shadows painted around the crater's inner edge. When these canvases were placed on the ground, the shadows had to be oriented to the position of the sun and turned periodically; but after pretesting at Farnborough by the RAF, they proved extremely convincing when seen from the air.

Turner's "bomb craters," pegged down flat and smooth on the runways, were no hindrance to an airfield's normal operations—though at first a number of startled pilots coming in for landings had to be reassured by their ground control officers. *"Don't worry about those, old boy—they're only to fool the Jerries."*

But make-believe bomb damage and blazes on nonexistent airfields didn't put an end to Colonel Turner's and Captain Board's inventiveness. As the air war and the bombings went on, Q lighting was extended to other targets such as aircraft plants, rail yards, and port facilities. The protection of fighter factories was especially urgent in view of Directive #17 from Hitler's Headquarters, dated August 1, 1940, which said in part:

> The [Luftwaffe] attacks are to be directed primarily against flying units, their ground installations, and their supply organizations, but also against the aircraft industry. . . .[9]

To win supremacy, the Germans would have to destroy the RAF not only in the sky and on the airfields, but in the factories as well. These plants were vital to Britain, and many were soon being protected by night decoys created by Colonel Turner's hardworking camoufleurs. Before a decoy was set up, an aerial reconnaissance was made; then the light pattern, situated a mile or two from the target, was designed to imitate the real objective as faithfully as possible. The decoy lights were dimmed and shaded to some degree, to simulate a hasty attempt at blackout discipline. The *real* installations were, of course, blacked out very thoroughly, and in many cases were camouflaged with nets and foliage. Special care was also taken to cover large glass factory windows from the outside, to prevent telltale reflections in the moonlight. While not as extensive as the airfield sites, the industrial decoys drew regular attacks by the *Luftflotten,* and British aircraft production was able to continue fairly undiminished.

By the end of August, pressure on the British defenders had grown even more severe. In his account of the campaign, historian Basil Collier wrote:

> Dowding's squadrons were still fighting magnificently, and were to go on fighting magnificently, but effectively their strength was dwindling. Faced with the more closely-knit formations the Luftwaffe was now using, they were losing more aircraft and pilots than they could afford,

> and . . . in spite of the widely-held theory that bombing aerodromes was a waste of time, the attacks on [Air Vice-Marshal] Park's sector stations were a powerful threat to their efficiency.[10]

The Luftwaffe had stepped up its daytime bombings as well as the night attacks, roaring over the English countryside in great waves; and the camoufleurs responded by planning a new decoy network that could be effective in daylight hours. These decoys, called "K" sites, would naturally have to be far more realistic and convincing than the simple night targets with their lights and Starfish fires.

To help build the K sites, Turner's crews were joined by set designers from British film studios; and scores of dummy airplanes simulating Hurricanes, Blenheims, Whitleys, Spitfires, and Wellingtons were manufactured by local display companies.[11] Because they were much more complex than the night decoys the K sites were fewer in number, but were prepared with extreme care. Each site displayed a runway, maintenance sheds, fuel and bomb dumps, trenches, roads, taxiing areas, and between eight and twelve dummy aircraft. Everything on the site was built of lumber, canvas, and fiberboard suitably painted, and a few real trucks and planes were sprinkled among the fakes to add authenticity. To provide some degree of genuine defense each "airfield" was also issued two heavy AA machine guns, and occasionally the gunners managed to bring down a raider coming in low to strafe or dive bomb.

Like the night sites, the day decoys were usually paired with operating fields. A site at Grange Town protected the RAF base at Thornaby, one at Thetford covered Honington, and a decoy at Taft Grange served as a shield for Hemswell. K sites were also provided for Marham, Debden, Acklington, Church Fenton, Lymington, Biggin Hill, and many other potential targets.

The original Q sites required only two trained men to

throw the necessary switches and trigger the "fire damage"; but the elaborate K decoys were each operated by a squad of twenty men directed by a flight sergeant, who reported to his "K-Area" pilot officer at the parent station. All of these men received special training in camouflage and the use of decoys at a secret school set up by Colonel Turner near Hook, in southern England.

To be convincing, a military decoy must show realistic activity as well as realistic objects, and the site crews worked hard, moving their dummy planes about the fields, faking truck and taxi tracks, rearranging the sham supply dumps, and periodically adding supplies as "new shipments" came in.

The men lavished much attention on their spurious bases, and sometimes became highly incensed when Luftwaffe planes flew in to attack, forgetting of course the purpose for which the sites were designed. A memorable exchange, recounted fondly by RAF Flight Lieutenant Robin A. Brown, took place over a K-Area field telephone between a flight sergeant and his pilot officer. The conversation went on against a loud background of exploding bombs and hammering machine guns:

> Flight Sgt. (agitated): Sir! We're being attacked!
> Pilot Officer: Splendid, Sergeant. Good show.
> Flight Sgt.: They're smashing the place to bits!
> Pilot Officer: Yes, excellent. Carry on.
> Flight Sgt.: But, sir—we need fighter cover! *They're wrecking my best decoys!*[12]

Later in the war, after the Allies won the upper hand and began carrying the air offensive to Axis territory, the Germans also used decoy targets to protect factories and airfields.

One of the most realistic of these was located at their Air Experimental Center in Rechlin, on the *Muritz Zee.*[13] The decoy, protecting a nearby bomber base, had a splendid runway, two large hangars, numerous small buildings, and scores of dummy He-111s and Ju-88s. This elaborate hoax was eventually uncovered by Allied photo interpreters, who in turn alerted American and British air commands.

Another enemy decoy, built in occupied Holland, led to a tale that has been told and retold ever since by veteran Allied pilots. The German "airfield," constructed with meticulous care, was made almost entirely of wood. There were wooden hangars, oil tanks, gun emplacements, trucks, and aircraft. The Germans took so long in building their wooden decoy that Allied photo experts had more than enough time to observe and report it. The day finally came when the decoy was finished, down to the last wooden plank. And early the following morning a lone RAF plane crossed the Channel, came in low, circled the field once, and dropped a large wooden bomb.[14]

Decoys and misdirectional lights were used at other times during the conflict, both by the Germans and by the Japanese. Many of these enemy ruses were effective but none reached the ambitious scale of Turner's conjury in Britain's crisis months of 1940.

By September of that year the RAF, though at great cost in men and planes, was still fighting stubbornly and successfully. The arrogant, pompous Goering, frustrated at his inability to crush British air power, abruptly changed his strategy and began reprisal bombings of the civilian population. And on September 5, 1940, Führer Headquarters officially directed the Luftwaffe to launch "harassing attacks by day and night on the inhabitants and air defenses of large British cities."[15] This led to a grim eight-month ordeal for the people of London and other communities, which came to be known as the "blitz." But for all the pointless death and destruction it caused, the shift in enemy tactics brought a respite to the hard-pressed British air units, and ended Hitler's hopes of ever invading England. As Churchill later reported, "The first German aim had been the destruction of our air power; the second was to break the spirit of the Londoner, or at least render uninhabitable the world's largest city. In these new purposes the enemy did not succeed."[16]

On September 17, the Nazi dictator was forced to order the postponement of Operation Sea Lion; then in mid-October he officially called it off, and the threat of German invasion was over. The RAF had stopped the Luftwaffe and won a victory which, in the prime minister's words, ". . . had been gained by the skill and daring of our pilots, by the excellence of our machines, and by their wonderful organization."[17]

In winning the air war, the fighters in the secret battle also contributed a significant share. During those critical months there were, according to the British Air Ministry, some *five hundred* decoy targets scattered throughout England, helping to divert enemy bombers from their goals and causing them to waste thousands of tons of explosives. The Q and K airfield sites alone, reported Colonel Charles W. Hinckle of the U.S. Department of Defense, drew over four hundred forty enemy raids, compared with about four hundred thirty raids on genuine RAF fields.[18] In short, while the major battles were being fought in the air, fully half the German bombing effort was being siphoned off harmlessly by the bizarre contrivances of Colonel Turner's Department.

In his story of the Battle of Britain, Churchill particularly mentioned the Starfish fire decoys which had, he noted, ". . . achieved remarkable results."[19] Given the uncertain touch-and-go nature of the air war and the narrow margins between failure and success, John Turner's piles of scrap lumber, roofing felt, and old rubber tires—blazing away in the darkness—had played their part well.

In that same account, Churchill added, "Away across the Atlantic the prolonged bombardment of London, and later of other cities and seaports, aroused a wave of sympathy in the United States, stronger than any ever felt before or since in the English-speaking world."[20]

The sympathy was tangible, and millions of Americans

shared deep feelings of kinship with the British; but there was also a strong isolationist sentiment and a mood of detachment. The United States, at the end of 1940, was physically untouched by the battle in Europe. Safe behind its great oceans, without hostile neighbors on its northern or southern borders, the country was still a sheltered oasis; and the roots of this national complacency—understandable in context—stretched back in history to earlier years, and far more naive ones.

5

"No Funds Available"

1940–1941

It took a lot in 1912 to scandalize the people of Paris, but Adolphe Messimy, then France's minister of war, managed to do it; and in the process to stir a huge tempest in an elegant military teapot.

On a visit to the Balkans to observe maneuvers, Messimy was impressed with the way the dull uniforms of the Bulgarian troops faded smoothly into the fields and forests, making them difficult targets. Soon after the Boer War the British army had adopted khaki clothing, and Messimy knew that the Germans were also shifting from Prussian blue uniforms to less conspicuous ones of field gray. Even the Russian troops of Czar Nicholas II had been issued tunics and trousers of grayish brown; but as for the French forces, they clung tightly to tradition. The *poilus* of 1912, historian Barbara Tuchman noted, still wore "the same blue coats, red kepi, and red trousers they had worn in 1830 when rifle fire carried only two hundred paces and when armies, fighting at these close quarters, had no need for concealment."[1]

Back in Paris, Messimy decided to change all that; the time had come to put French troops into good camouflage colors such as brown or gray-green. But when *M. le Ministre* announced his new plan there was a great national uproar. French newspapers printed shocked editorials, superannuated colonels wrote blistering letters, civic and veterans groups protested loudly. The glorious French army, dedicated to the concept of attack rather than defense, prided itself on its *élan vital,* its dash, its vividness. *"L'audace, toujours l'audace!"* had been Foch's dictum; what would happen to the troops if they were deprived of this spirit and went to war in drab, colorless uniforms? Government hearings were held and Messimy was bluntly overruled: the *pantalons rouges* were essential, the committee decided, to France's military image. Frustrated and unhappy, Messimy wrote in his diary that he dreaded the effects of this blind attachment to ". . . the most visible of all colors."[2]

War came and the minister's fears were realized. In August 1914 the French Third and Fourth Armies launched their first counterattack against the advancing Germans in the Ardennes, a sector which would make history again in a later war. General Joseph Joffre's proud officers, graduates of the elite St. Cyr Academy, charged into battle wearing handsome plumed shakos and brandishing glittering swords in their white-gloved hands. The Germans, better trained and using heavy machine guns—"innovative" weapons the French generals had scorned—melted in and out of the underbrush and the mists of the dense forest. They were hard to see and harder to fire at, but Joffre's battalions—clear targets in their blue coats and bright red trousers—were cut down by the thousands. The slaughter was devastating and the bodies of the confused *poilus* piled up everywhere.

Casualty lists make the best arguments, and Adolphe Messimy was vindicated by the silent corpses at Longwy and Rossignol. Soon afterward the French High Command, in some embarrassment, decided that loss of lives was more serious than loss of *chic,* and that less prominent uniforms for the troops might be advisable.

The clash between change and tradition is an ageless one; but in time of war change is generally accelerated. During World War I scores of new battlefield techniques were developed, and among these were many innovations in the art of camouflage. After the French disaster in the Ardennes, it was obvious enough that lives could be saved when troops wore uniforms that blended with the terrain. In addition, both sides in the war had begun to use front-line observers in balloons and small aircraft, so it became the practice to hide field artillery under foliage or camouflage netting. Bizarre color patterns were also being painted on Allied warships. The purpose of these unusual designs—known as "dazzle painting"—was to alter the apparent speed and direction of a vessel, and in that way to confuse range officers aboard German submarines.

Deceptions of this kind were adopted out of necessity, but most were considered of minor value and interest in them faded rather quickly. Between the two great wars the average military reaction to camouflage, in the American army as well as others, ranged from casual indifference to mild suspicion to active dislike. In combat operations, ambush and secrecy had always played an important role, but purely *protective* concealment was another matter. To action-oriented minds field camouflage had a taint of passivity or even timidness and, like the loss of the vivid *pantalons rouges,* was looked on as somehow "bad for morale."

Camouflage also violated a principle dear to the pipe-clayed heart of every staff officer: it went against the military obsession with neatness, spit-and-polish, drill-field regularity. Troops in garrison, as any veteran knows, have always lived by the numbers—their equipment spotless, barracks areas swept clean, lawns neatly manicured, supply and shelter tents lined up with perfect precision. Camouflage on the other hand was considered careless and "undisciplined." It made a virtue of dispersion, irregularity, and improvisation; and like modern ecologists, camoufleurs tried to blend with the natural landscape instead of exploiting and dominating it. All of which (so it was thought) would greatly complicate the practical problems of military logistics and organization.

One major factor would soon dramatically change the very nature and meaning of tactical concealment, and that was the growth of aerial photography; but before Pearl Harbor this wasn't yet of great importance, and attitudes in the American army were like those elsewhere.

Camouflage training, the responsibility of the U.S. Corps of Engineers, was one of the stepchildren of the services. Manuals and materials were dutifully prepared and experiments were conducted, but there was little real interest in the subject and even less money on hand to pursue it. At that point, America's armed forces were living on small budgets, the nation felt safe behind its ocean barriers, and war seemed far away. Sir John Dill, Britain's military representative in Washington, told his superior, General Alan Brooke, "This country is the most highly organized for peace you can imagine."[3] And Eisenhower, recalling those years, wrote that ". . . the United States Army mirrored the attitudes of the American people. . . . The mass of officers and men lacked any sense of urgency." As for weapons, he continued, the standard Springfield rifle ". . . was outmoded; there was no dependable defense against a modern tank or plane; troops carried wooden models of mortars and machine guns. . . . Equipment of all sorts was lacking and much of that in use had been originally produced for the . . . army of World War I."[4]

Around this fragile American cocoon the crises kept mounting. Hitler now dominated Europe, his U-boats prowled the Atlantic almost at will, the Wehrmacht divisions were massing to invade Russia, Rommel's desert tank battalions were growing in strength; and on the other side of the world the country's relationships with an expansionist Japan were deteriorating rapidly.

But despite the general complacency there were many

American leaders, both military and civilian, who became alarmed at the dangerous drift and wanted to correct it. Among them was Brigadier General Thomas M. Robins.

Tom Robins, assistant chief of engineers, a strong-minded man with thirty-six years' experience as an engineer officer, had a great many problems at the time, and one had to do with airfields. In August 1940, the construction of U.S. air bases and certain other installations was transferred from the Quartermaster General's Office to the Corps of Engineers. Every U.S. Air Corps base, in those casual years, was a highly inviting target; the installations sprawled carelessly over the landscape and could be seen for miles—open and vulnerable to enemy bombing attacks. Here, the Engineer Camouflage Board thought, was a chance to put theory into practice, and Robins agreed. Worried about the future, he fired off a memo to the War Department urging the immediate camouflage of all aviation bases in critical coastal areas. His memo stated:

> Modern air attack techniques, as demonstrated by European conditions, clearly indicates that concealment and camouflage of airfields . . . is of fundamental importance for those installations which are so located as to be in danger of aerial attack.[5]

Headquarters in downtown Washington (the Pentagon building hadn't yet been completed) asked for more details; but when Robins estimated that planning would cost in the neighborhood of $700,000 his request, according to the Office of Military History, "met with prolonged silence on the part of the General Staff." Today that amount would barely pay for the tail assembly of a single jet bomber, but in 1940 it was a sum of vast proportions; and Robins, very disappointed, had to inform the Board that ". . . the War Department apparently does not consider camouflage of fields important enough to justify the additional expense involved."

But the general was a stubborn man with the zeal of a

Messimy; and instead of dropping the matter, he and his District Engineers kept up their pressure. Appeal after appeal was filed, urgently asking reconsideration, and at the end of December the camoufleurs were given permission to try their concepts on *one* newly planned air base—Bradley Field, near Windsor Locks, Connecticut. The funds for the project were approved on the basis that it was strictly experimental.

Windsor Locks, some eight minutes' flying time from the Connecticut coastline, was in tobacco-growing country, a section of sprawling farms, colorful brick-red barns, and large green-textured fields. In designing the new base the District Engineer, Lieutenant Colonel John Bragdon, worked closely and enthusiastically with Robins and the Camouflage Board. On traditional fields "formal balance" was the chief goal, but Bragdon ignored all that. Instead of lining up his buildings in neat, symmetrical patterns, he dispersed them throughout the area and then had them painted a burnt-sienna tone to match nearby farm structures. The barracks were butted end-to-end with a common roof to resemble long tobacco sheds, and painted appropriately. Aircraft fuel tanks were put underground. Natural foliage was left intact wherever possible. Existing paths and roads were preserved, and new roads were allowed to follow normal ground contours instead of slashing across the landscape like great directional arrows.

The three long runways presented a challenge, and here the designers had to use restraint. "If you hide them *too* well," an Operations officer complained, "our own pilots won't be able to find them." Even with this injunction, a degree of illusion was possible. Roads, instead of stopping dead at the edges of runway paving, were continued across in matching paint colors. Here and there a section of green or brown field was extended—again with textured paint—right over the concrete landing strips, to break up their uncompromising lines.

"Bragdon," the OMH report states, "spared no effort to make Bradley Field invisible from the air." Test photographs, taken later, showed that his attempt was highly successful. It was difficult on the aerial photos to isolate the base proper from the surrounding countryside without several minutes of careful study, which was more time than would be available to any pilot or bombardier coming over in a fast-moving enemy plane.

The ingenious work at Windsor Locks showed what could be done to protect American targets; but it remained an isolated experiment, and despite new appeals by the Camouflage Board no further steps were taken.

During those frustrating months, Robins had a number of allies—men like Lieutenant Colonel Homer St. Gaudens and Major John F. Ohmer—who tried along with him to break through the walls of official indifference. St. Gaudens, son of the illustrious American sculptor, Augustus St. Gaudens, had served overseas as a camouflage officer in World War I. Between the wars he gained distinction as an art historian; then back in uniform again, and attached to the Corps of Engineers Construction Office, he continued to pioneer and proselytize. During the Battle of Britain and the subsequent London blitz, St. Gaudens made an inspection tour of England and filed detailed reports describing the brilliant camouflage ideas being developed there.

In Britain, the construction of decoys was handled by Colonel Turner's Department, but all other types of camouflage came under the Ministry of Home Security at Leamington. The Leamington projects dealt with the concealment of targets such as assembly plants, munitions works, supply depots, and aluminum factories. One installation in the Midlands, cited by St. Gaudens, involved a Fighter Command center and vast military storage area that stretched across thirty acres of open countryside.[6] To hide all this, the British built a huge canopy of fake "farmland" over it, complete with barns, sheds, country roads, and orchards; and at one spot there was even a wagon containing a realistic load of fertilizer.

St. Gaudens also observed British smoke-screen methods, which were then being widely used as a protective measure.

His report on this pointedly commented, "No factory protected by camouflage smoke screens has yet been bombed."[7]

Major John Ohmer, a small, peppery, enthusiastic officer, was attached at that time to the Camouflage Board at Fort Belvoir, Virginia. Like Homer St. Gaudens, Ohmer campaigned for the protective cover of critical American targets, and among other things prepared a detailed plan for camouflaging Wheeler Field in Hawaii. The plan, which wasn't adopted, had been cost-estimated at $56,210, a trivial sum by today's standards, and one which, in view of later events, might have saved scores of lives and many thousands of dollars' worth of needed planes and equipment.[8]

In spite of the general lack of support and funds, the Camouflage Board and its prophets kept doggedly promoting their cause; and some experiments were conducted on a modest scale at bases such as Maxwell Field, Alabama; Barksdale Field, South Carolina; Eglin Field, Florida; and Langley Field and Fort Eustis, Virginia.[9] In 1940 the board also negotiated with the Goodyear Rubber Company to manufacture rubber decoy aircraft such as were being successfully used in England. Goodyear produced several outstanding samples that could be duplicated in large numbers at a cost of $1,000 per "plane," and again there was polite interest but no follow-up.

On July 12, 1941—with Pearl Harbor only five months away—a memo from Lieutenant General Walter Short at Hawaii's Fort Shafter to the Adjutant General's Office in Washington reported:

> There is definite need for camouflage treatment of airfields in the Hawaiian Department. Up to this time no camouflage treatment has been undertaken on any air base in this department.[10]

Again no action resulted from his appeal, and everywhere the camoufleurs met with the same regretful litany: "No funds available."

Meanwhile, the tenuous years of peace for America were coming to an end. By spring of 1941, Hanson Baldwin wrote, "Japan and the United States were clearly committed to a collision course."[11] The Japanese, led by the belligerent Hideki Tojo, had vast imperialist plans for greater East Asia, and were actively preparing for war.

Hitler and his foreign minister, Joachim von Ribbentrop, encouraged them in these designs. The German dictator had great contempt for America's military capabilities, but was fully aware of the country's support for Great Britain and intended to act against the U.S. after he had disposed of Russia. As early as the summer of 1940, a top-secret memorandum prepared for the German General Staff stated, "The Fuehrer is at present occupied with the question of occupation of the Atlantic Islands [i.e., the Azores and Canaries] in order to operate long-range bombers from there against the U.S.A." [12] The Nazi dictator's military appetite was limitless. In the final analysis America, as Shirer pointed out, ". . . stood in the way of Hitler's grandiose plans for world conquest and the dividing up of the planet among the Tripartite powers. . . . The American Republic, he saw, would have to be dealt with eventually and, as he said, 'severely.' "[13]

The reality of this threat was becoming more and more apparent to Franklin Roosevelt, his advisors, and many other Americans, but awareness was one thing and mobilization another. The country's military power at that point was limited, and what did exist was diffused and vulnerable.

Ironically, John Bragdon's work at Bradley Field in Connecticut had been highly approved but because of priorities and limited funds no new projects were generated. On the day Pearl Harbor was attacked, Bradley remained the *only* U.S. military airfield properly protected by camouflage and dispersion. All other bases, including the crucial ones in Hawaii, were wide open to sudden destruction.

6

Out of the Cocoon

1941

They dove out of the sun over Oahu on Sunday morning, December 7, 1941, shortly before 8:00 A.M.—fast Val bombers, Zero fighters, and Kate torpedo aircraft launched from carriers of a powerful strike force commanded by Admiral Chuichi Nagumo.[1] In the sudden attack on Pearl the Japanese planes—one hundred ninety in the first wave and some one hundred seventy in the second—achieved complete surprise, their combined bombs, bullets, and torpedoes destroying men, planes, and ships in a quick, crippling blow; and within minutes Air Commander Mitsuo Fuchida was able to radio the *"Tora!"* signal indicating sure victory.

Hickam Field, Wheeler and Bellows Fields, the Marine Corps station and the Naval air base at Kaneohe Bay were all hit badly; and in large measure the work of the attackers was made easier for them. No attempt had been made to conceal or disperse any of the American aircraft. For protection against sabotage the Air Corps squadrons were kept close together and in every case the planes, Baldwin reported, ". . . had been lined up in neat rows in the open, wingtip to wingtip."[2] It was, for Fuchida's pilots, like flying straight down a giant bowling alley, and hardly a bomb or bullet went to waste.

In the harbor anchorage off Ford's Island the great ships of the U.S. Pacific Fleet were lying in trim, precise pairs along "battleship row," offering an obliging target impossible to miss. The *California* was anchored at the seaward end, followed by the *Maryland* and the *Oklahoma* paired side by side; just aft of these were the *Tennessee* and *West Virginia,* and next in line came the *Arizona* and *Nevada.* On the north side of the island, other battleships were similarly lined up. There were no barrage balloons or torpedo nets to provide a degree of protection.

In his memoirs naval Captain Ellis M. Zacharias, then commanding the heavy cruiser *Salt Lake City,* quoted from the remarkable diary of an anonymous Japanese officer who led one of the raiding squadrons. The diary said in part:

> Below me lies the whole United States Pacific Fleet in a formation I would not have dared to dream in my most optimistic dreams. I have seen all German ships assembled in Kiel Harbor. I have also seen the French battleships off Brest. And finally, I have frequently seen our own warships assembled for review before the Tenno. But I have never seen ships, even in the deepest peace, anchor at a distance less than 500 to 1000 meters from each other . . . It is hard to comprehend.[3]

The surviving soldiers, sailors, and airmen at Oahu fought back with great tenacity and courage; but in less than two hours the raid that Baldwin called "perhaps the most successful surprise attack of all time" [4] was over, leaving 2,334 Americans dead and 1,360 wounded, the major part of the Pacific Fleet destroyed or neutralized, and most of the available aircraft wrecked or damaged.

In the United States, people reacted to the Japanese assault with shock and disbelief; then shock gave way to anger and anger to action. The bombs and torpedoes that wrecked the fleet at Pearl Harbor had also shattered America's sense of security, and psychologically there was a complete turnaround. On the following morning the *New York Times* reported:

> The news of the surprise attack fell like a bombshell on Washington. President Roosevelt immediately ordered the country and the Army and Navy onto a full war footing. He arranged at a White House conference last night to address a joint session of Congress at noon today, presumably to ask for declaration of a formal state of war.[5]

Great Britain, whose forces in the Far East had been struck simultaneously by the Japanese, joined with America—as she had earlier with Russia—in a mutual cause; and on December 11, completing the circle, Germany and Italy officially declared war on the United States.

In London, Churchill and the British leaders were appalled at these sudden new developments. At the same time the prime minister, as he later indicated in his memoirs, was relieved when he considered their long-range implications, and noted:

> How long the war would last or in what fashion it would end, no man could tell, nor did I at this moment care. . . . The British Empire, the Soviet Union, and now the United States, bound together with every scrap of their life and strength, were, according to my lights, twice or even thrice the force of their antagonists. . . . Many disasters, immeasurable cost and tribulation lay ahead, but there was no more doubt about the end.[6]

Under pressure of the emergency, national differences in America vanished, the country united, and military leaders responded with speed and resilience. Among many other moves, General Henry "Hap" Arnold, commanding the AAF, ordered immediate "camouflage and protective revetments at all stations within the air frontiers, and . . . additional runways and auxiliary fields to permit wider dispersal of planes."[7] John Bragdon's classic camouflage plan for Bradley Field—the interesting experiment—was quickly dusted off and distributed throughout the Air Command, to serve as a future model.

The total absence of protective camouflage at Pearl Harbor was one factor that stood out glaringly in the endless aftermath of charges, investigations, and recriminations. Obviously, the use of such measures—the cover and dispersion, for example, of ships and planes—would not in any way have prevented the long-planned Japanese attack, but it would clearly have blunted its effects; and many American lives might have been saved just as French lives could have been saved at Rossignol and Longwy twenty-seven years earlier. But the violent message of that particular morning was finally grasped and understood, and would be acted on later with great effectiveness.

On January 6, 1942, in his State of the Union address to Congress, Roosevelt mirrored the country's new mood. The president called urgently for a rapid, all-out, unprecedented production of planes, tanks, antiaircraft guns, and other military equipment "to the utmost limit of our national capacity."[8]

America—at least for the duration—was out of the cocoon.

7

Disguise and Divert

1942–1943

If you flew in from the sea at Point Conception in 1942, followed the California coastline to Ventura, then turned inland over the Santa Susana foothills, you would note many small communities spread out along your route. Continuing east toward Burbank, you would fly over another of these tranquil areas with its modest ranch homes, tidy lawns, pleasant green trees, cars parked in the driveways, occasional gardens, and here and there some laundry hanging in backyards to dry. From your vantage point overhead it would all look quite ordinary, and you would have no reason to suspect that hidden below the placid scene was an enormous factory where thousands of people were hard at work turning out P-38 fighter planes for the U.S. Air Force.

This huge plant, the Lockheed-Vega factory, had been carefully camouflaged as part of a War Department emergency program to disguise vital installations and divert possible Japanese bombing attacks; yet the hiding of targets such as Lockheed-Vega was just one strand of a vast American defense network that developed in the immediate wake of Pearl Harbor.

A popular series of ads, which appeared at that time, showed a plain white cigarette pack with a large red bullseye in the center, and below came the frenzied message: LUCKY STRIKE GREEN HAS GONE TO WAR!

By mid-1942, chromium and copper used in the manufacture of printers' inks were urgently needed by the armed forces—but a lot more than Lucky Strike's green package went to war as the country geared for a two-ocean conflict. Sir Edward Grey, who had served as Britain's ambassador to the U.S. in 1919, compared America to a mammoth boiler. "Once the fire is lighted under it," he had said to Churchill, "there is no limit to the power it can generate." And now the American boiler began to generate power unmatched by any nation in history.[1]

In his January 6 speech, Roosevelt had set production goals for 1942 at 60,000 planes, 45,000 tanks, and 20,000 anti-aircraft guns plus "a multitude of other implements of war,"[2] and for 1943 all these figures would be greatly increased. General Ewart Plank, then director of national defense, recalled that his office was ". . . just given a blank check"[3] as Congress appropriated billions of construction dollars for training camps, shipyards, fuel depots, air bases, harbor defenses, ordnance plants, bombing and gunnery ranges, medical facilities, tank arsenals, munitions factories, auxiliary bridges, rolling stock, and other war-related projects. At the same time General Marshall made plans for mobilizing an army of 3,600,000, a force which would ultimately grow to over 16,000,000 men and women in the country's various military branches.

Among other matters, camouflage discipline was rediscovered. Once given lip service, it was now built into the basic training of every infantry, artillery, aviation, armored and supply unit; while at Fort Belvoir, Virginia and the Naval Research Center at Anacostia, outside Washington, D.C., designers experimented with ingenious new techniques for concealing personnel, ships, tanks, and aircraft.

Camouflage, from the French *camoufler,* is defined by

Webster as "the disguising of an installation, vehicle, gun position or ship with paint, garnished nets, or foliage to reduce its visibility or conceal its actual nature or location from the enemy." And to this must obviously be added the disguising of fighting men as well.

Green and olive drab—good low-visibility colors—had long since become standard issue, and the prodigious mills of New England and the South were soon rolling out millions of yards of khaki and field-green cloth and canvas for uniforms, tents, blankets, bed rolls, web belts, and other field gear. Meanwhile sporting goods factories, which once produced such frivolities as tennis, volleyball, and badminton nets, began to turn out thousands of camouflage nets for home defense and use overseas.

America's inexperienced troops had to be taught not only how to use their new weapons, tanks, and planes, but also how to protect themselves in combat. Camouflage training in the East was centered at Fort Belvoir, headquarters of the Corps of Engineers. In the Midwest, a large school was established at Jefferson Barracks outside St. Louis, Missouri. Training on the West Coast was handled at Hamilton Field, San Francisco, and later at March Field near Riverside, California. Smaller schools were scattered at many other bases.

The facility at Jefferson Barracks was directed by Major (later Lieutenant Colonel) William Pahlmann, one of America's leading interior designers. Prior to the war and after it, Pahlmann became a trailblazer whose concepts of interior and industrial design influenced the decorative arts in both Europe and the U.S. During the 1950s, he also produced the successful syndicated newspaper column, "A Matter of Taste." Geographically, the major's new school was in an important location. Hundreds of thousands of troops slated eventually to head overseas from either coast were funneled through Jefferson for basic combat training, which included the techniques of camouflage. Pahlmann's staff drew many notables from the world of the arts, among them Julian Harris, the sculptor; Gardiner Cox, the painter; Claude Herndon, designer of fabrics; Nathaniel Saltonstall and Hampden Robb, architects; and Donald Oenslager, the stage designer who created settings for such plays as *The Emperor Jones, The Man Who Came To Dinner, Stage Door, My Sister Eileen,* and Archibald MacLeish's *J.B.*

In a recent interview, Pahlmann, now graying and courtly, recalled those hectic days with affection. "At that point, early in '42, anyone in the arts was invited to apply for camouflage work—and we all rushed out there very green; not even knowing how to salute."[4]

Under the major's direction a unique school grew that would train GIs moving through Jefferson Barracks at the rate of 20,000 every few months. Among other things, the camoufleurs created a huge demonstration area depicting every important form of field deception. Camouflage control was heavily stressed, and a large sign at the entrance to the area warned:

> *DO NOT STEP OUTSIDE THESE PATHS!*
>
> They are wired to prevent you from making and leaving a mark for the enemy to see and to photograph. When you stay within a wired path, you are practicing a very important form of camouflage discipline.

As part of their demonstration, Pahlmann's crews built a complete French village equipped with realistic *son et lumière* effects. In it were scores of camouflaged strong points, dugouts, and sniper posts, as well as ingenious booby traps similar to those the trainees might encounter in combat. The booby traps were rigged with nearly invisible wiring to carelessly dropped enemy weapons, abandoned crates of ammunition and food, cases of French wine, and a variety of other "souvenirs." Moving any of these tempting items would trigger a satisfyingly loud (but harmless) explosion.

The demonstration area was run with such theatrical flair

that VIPs passing through Jefferson invariably asked to visit it; and later the civilian population of St. Louis was invited out on occasional Sundays to see a carefully edited version of the unusual display.

But Pahlmann's trainees were far from passive spectators. "Every man who went through there," he said, "had to work, and work hard. He actually became part of his own training. We taught the men how to protect themselves individually, how to make good use of their terrain. We kept emphasizing that camouflage meant fooling the eye—not only covering things up, but *diverting* the enemy's attention." Leafing through a stack of treasured wartime photographs he added, "Our staff was made up of artists, architects, designers—people who had a feel for color, for shape. Those were the tools we worked with."

The intensive training at schools such as Pahlmann's was only one aspect of America's new camouflage consciousness. With the bombings of London, Manchester, Birmingham, and Coventry in mind—plus the traumatic lessons of Oahu—the country adopted many protective measures. Antiaircraft defenses were strengthened. "Dimouts" were instituted on both coasts to cut down night glare. Blackouts and air raid drills became a regular occurrence in large coastal cities. Directors of war plants flooded Washington with requests for help and information on how to protect their factories. There was so much interest in this particular area that civilian training courses were organized in New York City—at the Cooper Union school and Pratt Institute—devoted entirely to techniques of industrial camouflage.

In addition to this, and in keeping with "Hap" Arnold's earlier AAF directive, the protection of vital U.S. airfields was given high military priority.

At Richmond Air Base in Virginia, a key installation, camouflage was supervised by the noted theatrical designer, Major Jo Mielziner. The major, one of Pahlmann's early associates at Jefferson, will be remembered for stage-designing such Broadway masterpieces as *Strange Interlude, Winterset, The Glass Menagerie, Death of a Salesman, A Streetcar Named Desire,* and *South Pacific.* Working with him at Richmond were talented craftsmen and artists including George Gray the muralist, and Scott Johnston, who later became a prominent illustrator of children's books. One artist on Mielziner's project was particularly adept at creating three-dimensional effects. In his spare time he painted a large *trompe l'oeil* "bulletin board" covered with thumb-tacked notices, an Air Force shoulder patch, a crumpled love letter, several photographs, and a set of metal GI dog tags. The painting was hung on a wall at headquarters and was so realistic that unsuspecting visitors would reach out to touch the items before realizing that they faced a flat painted surface.

At Richmond, as at Bradley Field, the base was designed to blend with surrounding civilian areas. The long low troop barracks were each painted in two pastel colors divided down the center, to resemble semidetached family dwellings. False chimneys for nonexistent fireplaces, small awnings, white picket fences, and a variety of shrubs and "front lawn" treatments added to the make-believe. All military vehicles were parked under trees, carports, or raised camouflage nets called flat-tops. Civilian cars were permitted on the base, but instead of being crowded into huge parking lots they were allowed to park at random along the many streets of this fictitious "suburban community."

For better protection, aircraft were dispersed and individually concealed, and hangar shapes were broken up with carefully placed nets. Mielziner had the field's runways coated with textured asphalt to reduce glare, then spray-painted them in patterns of green, brown, and yellow ocher to blend into the local countryside. By applying lighter and darker shades of color at certain points, a three-dimensional foliage effect was also achieved.

As expected, there were anguished cries from G-3 officers

who worried that the camouflage would confuse Air Force pilots attempting to land at the base. But Mielziner, a man of tact and diplomacy, was able to reassure them; and the record shows that no flier was ever lost over Richmond. At least, the camoufleurs claim, not permanently.

Years later, referring to his war work, the designer said, "On the stage my job was to make people grasp a situation as quickly as possible. In camouflage, my job was to keep them from grasping it at all."[5] Mielziner's views paralleled those of General Sir Archibald Wavell, one of the founders of the London Controlling Section. In a 1940 memo to the British Chiefs of Staff, Wavell had pointed out that:

> . . . the elementary principle of all deception is to attract the enemy's attention to what you wish him to see and to distract his attention from what you do not wish him to see.

It was by such methods, he added, echoing Houdini, ". . . that the skillful conjuror obtains his results."[6]

At this point it should again be noted that the conjury of warfare can be used either offensively or defensively. In the Fortitude hoax and other Allied deceptions, camouflage became in a sense an active combat weapon: The decoys and illusions helped greatly to mislead the Germans as part of a tactical operation. On the domestic fronts in America and England, camouflage was employed defensively: to conceal airfields, factories, and other installations and protect them in some measure from aerial attack. Many techniques and materials were available for this, such as nets garlanded with colored burlap strips, unusual paint treatments, natural foliage cover, and plywood or canvas structures that could disguise conventional forms and patterns. The aim of such concealment wasn't to hide a target perfectly, but simply to baffle and disorient an enemy flier *for the few critical moments of his bombing run.*

This goal was a feasible one. In the early years of the war, before the advent of radio-directional aids, aerial bombing was largely a visual operation. A bombardier in a Heinkel 111 or a Mitsubishi G4M traveled about three hundred miles per hour at an average height of fifteen to twenty thousand feet. At that speed and altitude, the target had to be identified while the plane was still five miles away; and to allow for trajectory, the bombs had to be released two or three miles ahead of ground zero. Which meant that as the plane hurtled toward its goal the bombardier had roughly thirty seconds in which to identify his target, adjust his range mechanism, and release his explosives. Added to that was the possibility of fighter plane interference, enemy flak, and poor visibility—all of which would hinder the raider's chances and greatly improve those of the camoufleur.

Given these difficult conditions a slight error by the bombardier could—and often did—result in bombs falling thousands of yards from the target; and it was the job of the military conjurors to make such errors happen as often as possible.

In early 1942, with America on the defensive, the dual goal of camouflage was to "disguise and divert," but nowhere was this need more serious than on the country's West Coast. The shock of Pearl Harbor had left its greatest mark on states bordering the Pacific, and for weeks afterward there were many rumors of impending attacks by enemy submarines and bombers. Frenzied newspaper accounts and scare headlines also appeared, adding to the alarm and confusion. These fears were, of course, understandable, and the Office of Military History reported:

> In the first two weeks of war it seemed more than conceivable that the Japanese could invade the coast in strength, and until June 1942 there appeared to be a really serious threat of attacks by a Japanese carrier striking force. These calculated apprehensions were fanned in the first

few days of war by a series of false reports of Japanese ships and planes on the very doorsteps of the Pacific states.[7]

In actuality Japan's threat to the mainland proved to be negligible. During February, two lone Japanese submarines did reach the West Coast.[8] One patrolled rather halfheartedly off San Francisco without any encounters and finally headed home. The other cruised near San Diego, and on the night of February 23, it surfaced off Santa Barbara and lobbed some shells at an oil installation. The damage was very minor and the submarine presently departed, but at that point everything hung in the balance: The emergency was real, the West Coast seemed dangerously vulnerable, and defensive measures of all kinds were urgently needed.

Camouflage training for the Western Defense Command was centered initially at Hamilton Field north of San Francisco, and then shifted to March Field, a large base on the fringe of the Mojave Desert some forty miles east of Los Angeles. The school at March was directed by John Ohmer—the same officer, now a colonel—who had pioneered during the prewar years with Robins and St. Gaudens. With him at the base was an Engineer Camouflage Battalion, responsible for additional training and experimentation.

Ohmer's work was made easier by the proximity of Hollywood with its huge pool of gifted moviemakers; and from M-G-M, Warner Bros., Universal Pictures, the Disney Studios, Twentieth Century-Fox, and other companies came a stream of volunteers and draftees—among them art directors, scenic designers, painters, animators, landscape artists, lighting experts, carpenters, and prop men. They ranged in experience from veteran art director Gabriel Scognamillo to neophyte Harry Horner, a protégé of Max Reinhardt who went on later to art-design outstanding films such as *The Hustler, The Heiress,* and *Born Yesterday.*

"None of those people," a personnel officer confided, "had the slightest bit of military know-how, but they sure made up for it with their energy."

Deluged by this flood of talent, the prosaic air base was soon awash with creativity, and its outlying sections resembled the back lot of a busy film studio. At March, it was hard to tell where reality ended and fantasy began. A casual visitor, wandering through the experimental area, might come upon hollowed-out tree trunks with ladders inside for artillery spotting, haystacks that split open to reveal machine guns, collapsible shacks that could disgorge a thirty-ton tank in the blink of an eye.

In one spot a huge craggy rock rose from the flat, sun-baked ground. At a signal the front of the rock swept aside and a P-51 Mustang fighter came roaring out, while simultaneously the top of the crag flew back to reveal an antiaircraft gun with its crew. At another point, triple rows of heavy concrete tank obstacles called "dragon's teeth" stretched across the terrain. Closer inspection would show that the ponderous obstacles were flimsy shells made of plaster. A third site offered the serenity of a small farm complete with barns, outbuildings, a silo and a pickup truck. The pastoral setting—created of two-by-fours and painted canvas—covered a large ammunition dump, and gained added reality when the camoufleurs persuaded a local farmer to graze his cows near the buildings.

One area of the base was devoted to experiments with decoy planes and gliders. Field camouflage depends heavily on improvisation, and the decoy aircraft reflected this. They were deliberately constructed of odd and unlikely materials such as canvas scraps, foliage, cardboard ration boxes, burlap on chicken wire, discarded packing crates and flattened tin cans. None of these makeshifts could, of course, withstand normal inspection, but when seen hastily from a speeding airplane several miles overhead, they proved surprisingly effective. The camoufleurs made fake runways for their "planes" by the controlled burning of grassy strips. They also tested new ways of building dummy tanks, trucks, bridges, searchlights, gun batteries, and even radar stations.

At Ohmer's school, scores of air force officers slated for overseas posts were trained in camouflage principles, methods, materials, the use of decoys, and the basics of aerial photography. The Engineers were also responsible for mass training hundreds of thousands of airmen at Fourth Air Force stations all over the Western Command. This crash program was directed by Major Walter Krotee, who organized a fleet of mobile schools that traveled from San Diego to Seattle, stopping for a few days of strenuous training at every airfield en route. Each cadre, consisting of ten or twelve camoufleurs, toured in its own mini-convoy complete with equipment, nets, displays, and collapsible "stage sets," holding classes and giving hundreds of camouflage demonstrations, which achieved a high level of showmanship. Lieutenant Harold Thresher, a commercial artist who took part in Krotee's unique marathon, said "After two months on the road, we began to feel like actors on an oldtime vaudeville circuit."

One element of the traveling demonstrations involved the concealment of well camouflaged "enemy snipers" in a large open field. Onlookers standing along one edge of the field were invited to locate the hidden men, and invariably "discovered" three or four who had been deliberately left detectable. The observers, searching intently, finally completed the hunt and congratulated themselves on their sharp vision; at which point a signal was given and an additional seventy snipers, with rifles ready, emerged from hollow tree trunks, fake shrubbery and well-hidden foxholes. Many were literally within a few yards of the onlookers, and the dramatic impact was considerable.

This quality of theater and stagecraft crept into almost all wartime camouflage, and the artists, designers, painters, and builders obviously enjoyed their assignments. But they also took them seriously: The goal of their work, as they knew, was to save as many Allied lives as possible, and in the contingencies of the war that goal was often realized.

U.S. camouflage projects in those months went far beyond training and experimentation.[9] On January 8, 1942, the War Department had ordered Western Defense Command to carry out strong "passive defense measures" for all vital installations within three hundred miles of the Pacific Coast, plus all those in Alaska. Accordingly the COE's District Engineers, working with company personnel and private contractors, began camouflaging industrial plants, aircraft factories, and fuel storage areas, as well as key air stations. In the next critical months, thirty-four airfields were given some form of protection, from simple color and paint "tone-downs" to runway texturing, foliage replanting and structural camouflage cover. Among these were Paine and McChord Fields in Washington; the Portland and Salem Air Bases in Oregon; and Mills, Salinas, and Hammer Fields in California.

But an even greater need was to conceal key war factories and assembly plants which, at that time, were looked on as highly inviting and vulnerable targets. The largest bombers in World War I carried bomb loads of only one thousand pounds, but by 1942 their lethal capacity had risen to over six tons per plane; and to protect American industry from that kind of destruction, the Pentagon earmarked funds that eventually totaled well over twenty million dollars.[10]

The Walker Construction Company, a California firm that played a key role in this program, had designed and built structures such as the Richfield Building, the headquarters of Southern California Edison, and the Douglas Aircraft plant in Long Beach. Now the firm was called on to help disguise the very structures it had created: to hide vast storage yards under fabricated trees and turn large factories into innocent pastoral landscapes. It was a prodigious job and H. M. Walker, the company's director, recalled that ". . . the Army was fortunate in being able to contact some of the finest special-effects men in the world right here in Hollywood, and these artists and craftsmen spent feverish months in preparation of plans and master designs to cover entire industrial areas."[11]

The Lockheed-Vega aircraft plant at Burbank, hidden beneath a complete southern California "suburb," was perhaps the most ambitious of these numerous disappearing acts.[12] With the cooperation of Lockheed's chief engineer, Hall Hibbard, the huge sprawling factory was covered by a giant canopy of chicken wire, scrim netting, and painted canvas. This vast umbrella was supported by a scaffolding of posts and cables, and the edges all sloped gently to the ground to melt into the landscape. Using endless gallons of paint, the artists carried local roads and streets up one side of the structure, over the top, and down the other slope; canvas houses were placed along the streets, and numerous trees and shrubs were "planted." The trees and shrubbery were constructed of molded chicken wire that had been treated with an adhesive; these forms were then covered with masses of chicken feathers, to give a leafy texture, and spray-painted in suitable tones of green. Facsimile cars were added here and there, as well as laundry lines and "victory gardens." Numerous air ducts provided ventilation for the plant's workers, and trapdoors led up to the canopy through the factory roof.

To be convincing, a simulation of this type must show normal signs of life and activity. The camoufleurs had access to the area through the trapdoors; moving about on hidden catwalks, they were able to do necessary maintenance work and to move the lightweight "parked cars" periodically. They were also very careful to put out and take in the regular Monday wash.

This unique deception was so extensive and well-constructed that when it was finally removed the dismantling alone cost the War Department some $200,000. The artisans who worked on Lockheed-Vega's camouflage gained valuable expertise; and in later years some of them used the same techniques in helping to design and build California's fabled Disneyland.

In Seattle, Washington, the twenty-six-acre Boeing Aircraft plant was also given an elaborate cover-up; here as at Burbank, a complete "town" was created over the plant proper with streets, driveways, trees, lawns and over fifty private "homes," all made of canvas and fiberboard. Similar but somewhat less spectacular disguises were created for other West Coast targets, among them the Douglas Aircraft plants in Santa Monica, Long Beach, and El Segundo; North American Aviation in El Segundo; the Vultee plant in Downey; Northrup in Hawthorne; and the Ryan and Consolidated Aircraft companies, both located in San Diego.

All this served a valuable purpose through most of 1942, then the pendulum began to swing. By November of that year, with U.S. Marines on Guadalcanal, the Russians holding at Stalingrad, and Rommel's troops falling back from El Alamein, the Allies' military fortunes began to improve rapidly. Added to this, the enormous growth of American air power and the U.S. Navy's defeat of the Japanese advance carrier force at Midway put a final end to fears of enemy bombers reaching the American mainland. The Navy's decisive victory at Midway, reported the OMH, ". . . virtually ended the threat of a serious attack on the West Coast," since Japan's remaining naval and carrier strength would be "needed to defend their earlier conquests."[13] There were, of course, years of heavy fighting still to come, but the continental U.S. was no longer in danger.

With this easing of pressure, camouflage measures on the home front naturally lost urgency and the work gradually came to an end. Later, some of the camouflage battalions were disbanded and their personnel reassigned to activities such as cartography and photo interpretation. Other Engineer units went overseas to deal with camouflage problems in the combat zones, where the need was critical.

In battle, with the Allies beginning to take the offensive everywhere, protective measures differed from those used on the home front, but the basic principles applied. Reporting from central Tunisia in 1943, where Stuka dive bombers were active, noted U.S. war correspondent Ernie Pyle wrote that ". . . camouflage became second nature. Near the front no one ever parked a jeep without putting it under a tree. If there were no trees, we left it on the shady side of a building or wall. If there was no cover at all we threw a camouflage net over it. As we neared the front we folded our windshield down over the hood and slipped a canvas cover over it so it wouldn't glint and attract a pilot's eyes."[14]

Under these conditions, whether it involved a lone rifleman, an infantry company, a command post, an armored column, a forward airstrip or a supply convoy, cover and concealment in all its variations—with live bombs, shells, and bullets flying—took on a new and very real meaning.

A U.S. Sherman tank, made of inflated rubber, convincingly simulates real armor when seen from the air. Scores of these rubber tanks were used by the Allies in the huge preinvasion deception, Operation Fortitude.

(COURTESY IMPERIAL WAR MUSEUM)

Inflatable trucks were also used in Fortitude, particularly in the Ramsgate-Hastings area, to reinforce "invasion build-up" ostensibly aimed at the Pas de Calais.

(COURTESY IMPERIAL WAR MUSEUM)

Dummy assault vessels such as this inflated LCT were moored along the Channel coast as part of the Allied Fortitude deceptions. (U.S. ARMY PHOTOGRAPH)

Fake landing barges built of canvas and lumber displayed in the Thames estuary and at numerous points on the Dover coastline to support the "Calais invasion" hoax. (COURTESY IMPERIAL WAR MUSEUM)

Above, British gun position disguised as a concrete abutment on Westminster Bridge (*see arrow*) in the center of London. *Below,* antitank artillery hidden inside fake petrol station on a coastal road in East Anglia. (COURTESY IMPERIAL WAR MUSEUM)

Genuine U.S. Army transport vehicles near Portsmouth, England, well camouflaged in preparation for D-day and the actual Normandy landings.

(U.S. ARMY PHOTOGRAPH)

Major William Pahlmann at Jefferson Barracks, St. Louis, modeling early version of U.S. Army's jungle camouflage suit.

(COLLECTION OF WILLIAM PAHLMANN)

Dummy tree stump designed to hide an observation or sniper post, constructed at the Jefferson Barracks camouflage demonstration site.

(COLLECTION OF WILLIAM PAHLMANN)

Camouflage on the Atlantic Wall: This German gun position at Tourlaville, near Cherbourg, was covered with an innocent-looking "barn roof," but it failed to deceive Allied photo interpreters or bombardiers.

(COURTESY U.S. AIR FORCE)

A Wehrmacht pillbox (*see arrow*) hidden in the base of an elegant château commanding the harbor at Le Havre.

(COURTESY IMPERIAL WAR MUSEUM)

Low aerial view of one section of the camouflage experimental area at March Field, California. Photo shows a number of decoy fighter planes and several trucks, all built of canvas and plywood. (COLLECTION OF THE AUTHOR)

"Butch," a dummy B-24 bomber built by camoufleurs and made entirely of wooden planks, near Indio, California. (COURTESY U.S. AIR FORCE)

Defensive legerdemain in California: an aerial view of the Lockheed-Vega aircraft factory in Burbank, hidden under a vast canopy of shrubs, lawns, and ranch houses. Camouflage extends to the upper margins of the photograph. In this view, the remarkable masquerade is about ninety percent complete, with a few finishing touches still remaining. (COURTESY LOCKHEED AIRCRAFT CORPORATION)

The Lockheed-Vega plant in early 1946 after the camouflage cover was removed. Comparison with above photo shows extent of the original deception. Note new aircraft test runway in foreground with faint traces of camouflage paint pattern still visible. (COURTESY LOCKHEED AIRCRAFT CORPORATION)

Part of the large parking lot at the Lockheed-Vega plant in Burbank. Cars have been screened from aerial view by a flat-top camouflage net.

(COURTESY LOCKHEED AIRCRAFT CORPORATION)

Free-form sculpture in the American desert: a tank destroyer under a draped camouflage net, on maneuvers near Mojave, California. (U.S. ARMY PHOTOGRAPH)

U.S. Army exercises near Shreveport, Louisiana, in 1941, produced this unusual trompe l'oeil. Two large pyramidal tents, serving as Second Army Headquarters, masquerade as a rubbish pile adjoining a small sheet-metal factory.

(U.S. ARMY PHOTOGRAPH)

U.S. photo interpreter preparing special damage report on munitions plant and arsenal at Weimar, Germany, after bombing by the Eighth Air Force. Arrow indicates stereo viewer used in studying photographs three-dimensionally.

(COURTESY U.S. AIR FORCE)

Aerial view of main German rocket testing area at Peenemünde, on the Baltic coast. This was the first photograph in which an actual V-2 rocket *(see arrow)* was detected by the Allies.

(COURTESY IMPERIAL WAR MUSEUM)

Same area following the massive RAF bombing raid in August of 1943. Damage to the site was so severe that German rocket experimentation was set back for many months.

(COURTESY IMPERIAL WAR MUSEUM)

The P-38 Lightning, stripped of excess weight, became the standard U.S. photo reconnaissance plane in the European theater. The recon version of this versatile aircraft was known as the F-5. (COURTESY U.S. AIR FORCE)

Aerial hide-and-seek: An RAF Meteor Mk-3 with mottled camouflage design blends almost perfectly with the landscape pattern below.
(COURTESY IMPERIAL WAR MUSEUM)

A Douglas A-20 airplane under a typical flat-top net at a field in Surrey, England. (COURTESY U.S. AIR FORCE)

Dazzle design on a U.S. P-51 fighter. This type of unusual aircraft treatment was experimental, and rarely used in actual combat. (COURTESY U.S. AIR FORCE)

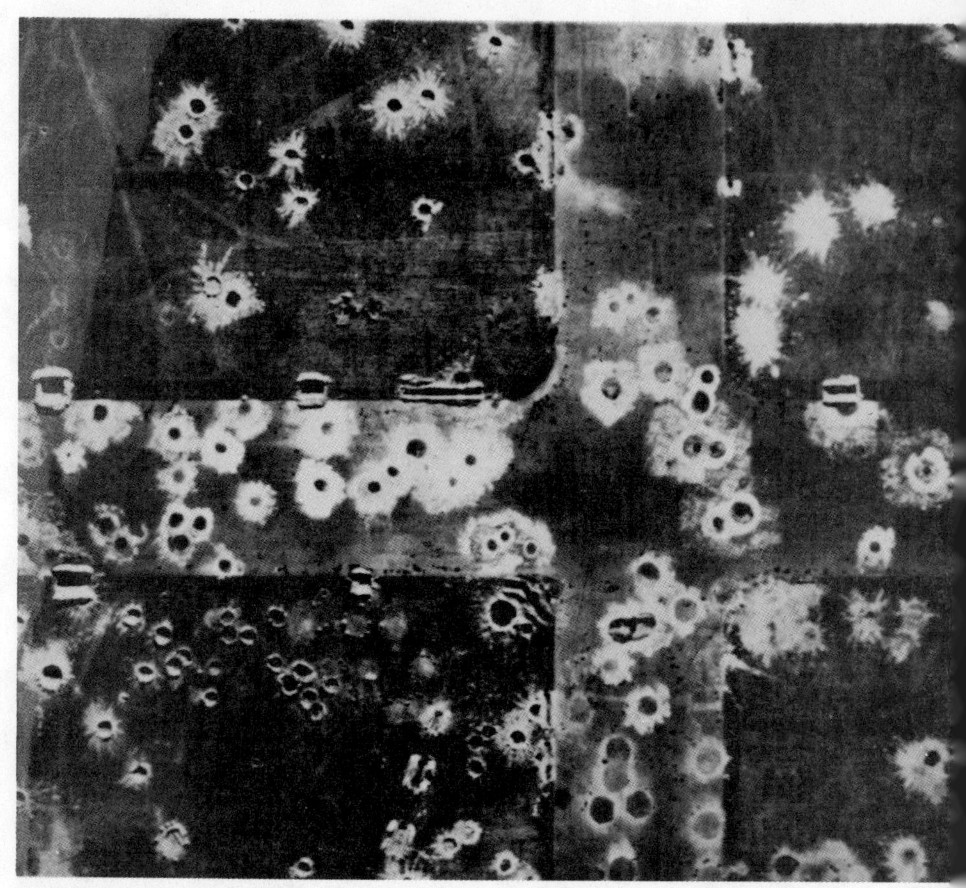

Visual sophistry by the Luftwaffe. Two runways at the Rheine airfield in northwest Germany had been heavily bombed by the Allies. Following this raid, German camoufleurs repaired the damage, then painted the filled areas to look as though the runways were still badly cratered. Three-dimensional stereo viewers enabled Allied PIs to spot this deception. (COURTESY IMPERIAL WAR MUSEUM)

Above, engineers of a U.S. camouflage battalion at Issy, France, conceal a large fuel storage tank within range of enemy artillery. *Below,* the final result.

(U.S. ARMY PHOTOGRAPHS)

Mass-produced dummy tanks of painted fiberboard mounted on jeep chassis were used in many of the deception schemes in North Africa.

(COURTESY IMPERIAL WAR MUSEUM)

Specially designed trailers such as this were pulled behind the dummy armor to create convincing "tank tracks." Large drums filled with water were mounted on the trailers to provide added weight. (COURTESY IMPERIAL WAR MUSEUM)

Aerial view of four-gun U.S. artillery battery, well camouflaged, in the outskirts of Castel di Casio, Italy. Note that on the two positions in upper part of photo, the deception is weakened by excessive tracks visible in the snow.

(U.S. ARMY PHOTOGRAPH)

Minturno, Italy: camouflage netting hung vertically to screen U.S. XI Corps supply road from ground observation. This type of concealment, used occasionally in World War II, was more prevalent in World War I. (U.S. ARMY PHOTOGRAPH)

Above, camouflage stresses inventive use of local materials. Here a squad of GIs near Poretta, in snow-covered northern Italy, dresses a net with spun glass obtained from a local factory. *Below,* a U.S. Fifth Army tank, wearing white paint patches and festooned with the same spun glass, achieves a high degree of protective cover. (U.S. ARMY PHOTOGRAPHS)

A hidden Japanese bunker in New Guinea after area was taken by Allied troops. The machine gun opening, also camouflaged, was positioned so defenders could attack GIs from the rear. (U.S. ARMY PHOTOGRAPH)

U.S. howitzer battery firing on enemy positions in the Solomon Islands. Note protective netting and mottled paint on gun barrels. (U.S. ARMY PHOTOGRAPH)

Workers at a small factory in Brisbane, Australia, weave strips of dyed canvas into camouflage nets for use in Pacific campaigns. (U.S. ARMY PHOTOGRAPH)

American infantryman modeling typical Japanese sniper's cape made of hemp fibers. (U.S. ARMY PHOTOGRAPH)

Misdirection at sea: dazzle painting on the cruiser U.S.S. *San Juan*. Such bizarre patterns, it was believed, would create course distortion and confuse range officers on enemy submarines. (COURTESY U.S. NAVY)

Dazzle design on British battleship H.M.S. *Rodney*. Royal Navy camoufleurs were particulary adept at this type of visual sophistry.

(COURTESY IMPERIAL WAR MUSEUM)

U.S.S. *New Orleans* in the Philippines area, camouflaged with foliage and garlanded nets to blend with nearby shoreline. (COURTESY U.S. NAVY)

Smoke-making equipment in action—part of the dense smoke screen laid down prior to Allied crossing of the Rhine River in March of 1945.

(COURTESY IMPERIAL WAR MUSEUM)

Below, the fine art of camouflaging water: England's Coventry Canal, shown here under a mantle of emulsified coal dust used to eliminate reflection on moonlit nights. Photo was taken four hours after the chemical coating was sprayed on.

(COURTESY IMPERIAL WAR MUSEUM)

8

Everything but the Ammunition

1942–1944

Cartoonist Dick Brown's sketch in *Yank,* the U.S. Army's wartime magazine, shows two irate fighter pilots in a jungle clearing, glowering at a benign, highly self-satisfied sergeant who stands clutching a large can of paint and a dripping brush. One of the baffled pilots is growling at him, "Okay, Rembrandt—where the hell's our planes?"

Far from home, on the battlefields of New Guinea and Tarawa, Tunisia and Anzio, Arnhem and the Hurtgen Forest, where Allied troops came under heavy enemy attack, the cover techniques learned in training were put to effective use. Conditions in all those areas varied greatly but the dangers remained the same; machine gun, mortar and sniper fire were ever-present, and even where the Allies had powerful air superiority there was no telling when enemy planes might suddenly roar down to strafe and bomb. Because of that, camouflage consciousness was high and the canons of disguise and divert were regularly followed.

The earliest Americans had used their own forms of camouflage. In 1564, Nicolas Le Challeux joined a French Huguenot colony in Florida and later wrote an account of the life of the natives. Le Challeux noted in his report:

> The Indians hunt deer in a way we have never seen before. They hide themselves in the skin of a very large deer which they have killed some time before. They place the animal's head upon their own head, looking through the eye holes as through a mask. In this disguise they approach the deer without frightening them. They choose the time when the animals come to drink at the river, shooting them easily with bow and arrow.[1]

The Florida Indians without access to Sun Tzu understood the nuances of deception; and hundreds of years later—in a grimmer and far more deadly context—the same ideas still applied.

Field concealment in World War II involved no magic ingredients of invisibility, but simply the optical legerdemain that resulted from breaking up familiar shapes, contours, lines, and colors which the enemy normally expected to see. Conversely, the *unexpected*—like Birnam wood approaching Dunsinane—had to be concealed from hostile eyes as simply and as efficiently as possible. Geoffrey Barkas, a camouflage expert who created many of the brilliant deceptions used by the British in North Africa, pointed out that, ". . . whatever additions man makes to the objects or marks on any patch of ground will inherently tend to be conspicuous if they run counter to the existing pattern, and inconspicuous if they continue the pattern or conform to it."[2]

Barkas's axiom was useful in combat. Infantry troops regularly practiced dispersion, and wherever possible took advantage of trees and natural foliage cover. The ubiquitous foxhole—camouflaged when time allowed—became the U.S. foot soldier's mobile home; and his distinctive round helmet often carried a scrap of netting festooned with twigs and branches to make it less obvious.

In the field, camouflage nets became standard tools for protection. Made of heavy twine, they ranged in size from roughly fifteen feet to forty feet square, and were garlanded with strips of burlap or cotton canvas (called "osnaburg"), which varied in color from tones of green to earth brown, loam and sand, depending on the local terrain. The colored strips were woven thickly at the center of the webbing and were gradually thinned out toward the edges. When one of these garlanded nets was thrown over a large object and properly draped it gave excellent cover, disguising hard shapes and diffusing the shadows: Seen from the air, a tent, truck, half-track, or artillery piece under a well-placed net was suddenly transformed into a grassy hill or a patch of innocent shrubbery. Many tanks and armored vehicles carried their own nets throughout the campaigns, as part of their regular equipment. Later, small-mesh or "shrimp" nets were issued that didn't require extra strip garlanding, but these were generally less effective.

In rear areas, texturing was sometimes applied to reduce surface glare on hangars, roofs, and runways. Texturing materials included gravel, redwood fibers, pine branches, rock and wood chips, excelsior, grass cuttings, and feathers.

The combatants also protected their tanks, half-tracks, and other armor by using bizarre color patterns. In Europe, armored vehicles were often painted with mottled designs of green and brown to help them blend into bushes and hedgerows. In the North African desert, tones of sand and ocher were used. Russian tanks for the winter campaigns were painted white. In Sicily, where the landscape is marked by rocky volcanic outcroppings and escarpments, one German panzer unit painted its armor with large gray stonelike patches. When the tanks were ranged against the side of an old crumbling stone wall, their *trompe l'oeil* proved highly effective. Again, the purpose of all this wasn't to make the vehicles "disappear," but simply to blend, blur, and disrupt their customary lines, and as a result make them harder to hit as targets.

Aircraft were treated in similar ways with countershading, much like the coloration of fish: the upper surfaces of many planes in World War II were painted in tones to blend with the ground, and under-carriages were colored light blue, to make them less conspicuous when viewed from below against the daytime sky. Some RAF reconnaissance planes were given a complete coating of pale bluish-green, and flying at high altitudes they became literally impossible to detect. Aircraft, which operated in darkness, such as the F-3 used by the U.S. Ninth Air Force for night photography, were painted jet black with a dull, nonreflecting finish to absorb the glare of moonlight.

Color also helped to protect individual soldiers. England's military contribution to uniforms had been the development of the color khaki at the beginning of the twentieth century; and variations were soon adopted by armies everywhere. In World War II all the combatants, Axis as well as Allied, put their troops into muted or mottled clothing to make them less conspicuous.

The basic U.S. field color—a blend of bronze and green—was the well-known olive drab. British soldiers in the North African desert wore sand-colored uniforms. Operating in snow, American, Norwegian, and Russian troops—later the Germans—used white capes and hoods. In the Pacific theater, with its elements of guerilla warfare, U.S. Marines wore special jungle suits and canvas helmet covers patterned to blend with the dense island foliage. Similar camouflage designs were later adopted by the British, also by the Germans who issued them to their panzer grenadiers and the elite Waffen SS. Parachutes for American airborne battalions, initially made of white nylon, were also imprinted with mottled green brown designs to make them less noticeable to enemy pilots flying over the drop zones.

The Camouflage Board at Fort Belvoir faced another big problem. Infantry troops in the field, no matter how well disciplined, often leave telltale signs, clues, and bits of debris. The sun reflecting on a gasoline can, a jeep windshield, or a shiny ration tin—or a few pieces of white laundry draped over a bush to dry—could become signal beacons for enemy planes scouting overhead. To lessen this kind of risk, low-visibility olive drab was soon extended to every conceivable item of G.I. supply and equipment, including shell casings, fuel containers, bedrolls, rain ponchos, wool gloves, handkerchiefs, K-ration boxes, even candy and chewing gum wrappers.

Sergeant Irwin Greenberg, who served with a paratroop combat team in Italy and later fought in the drive through southern France, recalls, "When I first went overseas, everything they issued to me was olive drab. Socks, underwear, blankets, towels—everything but the ammunition." [3]

Besides attempting to conceal and protect their own troops and installations in the field, both sides naturally tried to pierce the deceptions of the enemy; and combat intelligence teams attached to forward units worked continually to uncover enemy plans, tricks, and camouflage tactics.

As an odd sidelight to this, Sergeant William Manson, who served in Europe with the 82nd Airborne Division, deserves particular mention.[4] Bill Manson happens to be color blind, and during the war this gave him a surprising ability to locate concealed German positions. The sergeant's retinal singularity applies to the full blue-green spectrum, and he was able, when studying terrain hundreds of yards away, to spot camouflage nets and artificial foliage, which to other eyes appeared completely natural.

During the division's operations in France, Belgium, and Holland, Bill Manson's gratified company commander often used him on patrols to locate enemy machine gun nests and hidden 88 mm. gun emplacements. Manson himself is somewhat at a loss to explain his unusual gift. "At first," he said, "I thought I just had awfully good eyesight, but it was more than that. Actually, a color-blind person sees exactly what you see, but he can't tell which color is *which.* It helped me somehow to spot fake greenery. A gun battery under a painted net would show up clearly to me; I'd point it out to the others, but nobody else could see it until I described the exact bush, or the exact spot in the hedgerow."

Ex-Sergeant Manson, now a top executive with a New

York insurance company, recalls his role as a camouflage spy with fondness. "It wasn't anything official, and we only used it on the company level; but I must say it was the only time when being color blind was an advantage."

On a larger scale, under battlefield conditions, "fool the eye" came of tactical age. The art of camouflage had gained considerable status; and compared to the French commanders of 1914 who had clung so fiercely to the *pantalons rouges,* attitudes now shifted completely to the opposite end of the visual spectrum. But this great stress on tone, texture, and low visibility, on concealing "everything but the ammunition," was far from arbitrary: It was dictated in large measure by the growth of aerial reconnaissance, a development which changed the very nature of modern warfare and certainly the parameters of military deception.

During World War I, in Adolphe Messimy's era of attrition and trench fighting, camouflage—though useful against aerial spotters—was needed chiefly for horizontal protection; but with the coming of fighter-bombers and especially photographic planes, concealment became a vertical problem as well. Cover and deception had always been valuable military devices, and now the camoufleurs in both the front lines and rear echelons faced a new challenge; it was one thing to fool a human observer at ground level, but another matter entirely to outwit the cold, impersonal eye of an overhead camera. As the campaigns unfolded, this led to a curious percipient contest. The best aerial photographs in the world are of little value without trained people to interpret what they reveal. This automatically pitted the skills of the photo experts directly against the skills of the camoufleurs, who knew exactly what the interpreters *would try to find* on their photographs. The result was a visual chess game in which—depending on the particular operation—each side switched position, playing camera deceiver at one point and photo detective at the next.

A year before the outbreak of the conflict, Wehrmacht

General Werner von Fritsch had said, "The military organization that has the best aerial photo reconnaissance will win the next war."[5] The general's opinion can be debated, but the fact remains that in World War II the impact of aerial reconnaissance on tactics and planning became enormous; and there can be little doubt that, of all the war's lethal new weapons, one of the most dangerous was a simple, unassuming photograph.

9

"Let the Photos Speak . . ."

1941–1945

The aerial prints showed a drab cluster of abandoned French farm buildings, the roofs partially collapsed and the old stone walls scarred by shellfire and bomb damage. There were no signs of life anywhere. Major Driscoll, the unit commander, studied the photographs with vague uncertainty while a young lieutenant hovered over his shoulder. "I'm fairly sure of it, Major. It's either an advance headquarters or a GCI center. I think they left the damage there to throw us off the trail."

Driscoll picked up a small stereo device, placed it on two of the photographs, peered through and shifted the prints around, trying for the right visual effect. Under the twin magnifying lenses of the viewer (similar in design to an old-fashioned stereopticon), the farm buildings suddenly fused into three dimensions, almost as though he were hovering over them in a balloon, and now everything stood out with great clarity.

The Major frowned. "If you're right, the Germans are doing a good job. It looks deserted to me. Where are they keeping their transport?"

The lieutenant pointed to a patch of woods a hundred yards from the barn complex. "Their motor pool's in here, under these trees. The area's much too heavily wooded—it's not normal for this time of year. They probably added a lot of artificial foliage." He indicated a rutted dirt road skirting the far side of the woods. "At one point the road turns in under the trees—you can just barely see it—then a little farther on it comes out again. They doctored the gap in between to make it look like the road goes straight past the woods, but it doesn't—it's their access route."

He leaned over the desk and traced a finger along the print. "From the woods they walk right along this hedgerow into the main barn. There's probably a covered path there, or a tunnel underneath. Besides, it's the wrong place entirely for that kind of hedgerow."

George Driscoll continued to squint uncertainly through the stereo lenses. The interpreter's theory was feasible, but pinpointing camouflaged targets was a tricky business and they had to be careful. The photo unit had made a few embarrassing mistakes lately, which had been costly in wasted bombs and fuel, and the major was anxious to avoid another blunder.

"Haven't you got more to go on?" he asked. "What about defenses? If this is operational, there should be heavy guns somewhere."

The lieutenant pointed again. "They're probably in these wooden sheds, which would give them a good field of fire. And look at that haystack. Farmers in this part of the country don't stack hay that way; they build long low mounds, but this one is high and conical. It could be hiding antiaircraft." He put a second pair of photos on the desk alongside the first. "I checked a sortie flown ten days ago over the same area. Look at this, sir, the two wooden sheds were *added* since then. And compare the road coverage. The first one's fairly faint and hardly used, but since then it's been heavily traveled."

"Maybe by retreating Germans."

The interpreter persisted. "What about this clump of shrubbery? It *couldn't* have grown that much in ten days. I'm not sure, but I think I can see some radar equipment in there, possibly a 'Freya.' "

Driscoll studied the prints again. The new photographs had been taken at eight thousand feet, which was a relatively low altitude, but the day had been overcast and there were no clear shadows—unfortunate, since shadows often provided the best clues. He concentrated on the suspicious clump of shrubbery. Was there a faint glint of metallic edge in those bushes, a vague blur caused by a turning radar grid? Fretting over the stereo viewer, Driscoll recalled a comment by an RAF photo expert. "Don't strain too much to look for things," the veteran had said. "Just let the photos speak to you."

The half-hidden access road, the extra wooden sheds, the anomalous haystack, the shrubbery with its faint but tantalizing blur suddenly had something urgent to say. Driscoll swept the photographs together and handed them back to the interpreter. "Write it up," he said, "but don't put it through channels. Bring it back here. We'll send it direct to G-2 by air courier, and start worrying."

Within an hour the lieutenant's detailed report plus an annotated print was on its way, followed by a phone call from Driscoll to Operations recommending that the target be given priority; and early the next morning a flight of P-47s lifted from an airstrip in Normandy and headed for the cluster of barns with racks of bombs under their stubby wings.

The planes ran into unexpectedly heavy ground defenses over the target and German troops erupted everywhere, but they made their drop and got safely back to base. At the debriefing, the pilots reported that the attack had apparently caused considerable fire and damage. The adjoining patch of woods had also been strafed; the fliers heard heavy explosions and saw thick columns of black smoke billowing out. A visual reconnaissance flight that afternoon confirmed that heavy smoke, the kind made by burning fuel oil, was still pouring from the woods, and several days later a report from VII Army Corps brought additional news. Allied patrols in the area had taken some prisoners who told their interrogator that the "farm" had been an important Wehrmacht communications center, heavily camouflaged. According to the POWs, the bomb damage had been severe, much valuable radar and ground control equipment had been destroyed, and German operations throughout the sector had been badly disrupted.

When the report from VII Corps arrived, Driscoll made no special fuss. He simply dropped the teletype on the lieutenant's desk, tapped him approvingly on the shoulder, and hurried off to other and more pressing concerns. The interpreter, however, savored his triumph and congratulated himself at inordinate length. It can be noted further, since the author of this account was the young officer, that for a while he became insufferably smug.

Photo detection of this nature actually went on regularly as Allied armies fought their way across Europe from the beachheads of Normandy. Aerial reconnaissance or "recce" units, working with the advancing ground troops, moved forward with them mile after mile and provided valuable operational data. In combat, tank and infantry units needed information about enemy troop dispositions, the location of armor and heavy artillery, the general conditions of the terrain. Tactical air units needed information about the condition of bridges, roads and rail lines, the location of camouflaged targets, the movement of enemy supply convoys.

Most of this information came from overhead reconnaissance, a military "weapon" which had been tried experimentally in earlier campaigns. Balloon observers were first used by Napoleon's armies in Italy and Egypt, and later by General George McClellan in the American Civil War; but air intelligence as a practical technique had its real origins in World War I. In 1914, small scout planes were sent out so pilots could observe troop movements, spot artillery positions, and make sketches of enemy trench layouts. However, these missions, dependent on fallible human eyesight, weren't too successful;

the pilots didn't always see what they were supposed to see or remember what they saw afterwards. A camera, on the other hand, could see everything, remember everything, and report it all back with great precision. In addition, the prints brought back by the photographers could be examined at leisure and without enemy interference.

By mid-1915, both the Allies and the Central Powers were experimenting with aerial cameras, but there were still many problems. The first cameras were heavy brass-bound models operated by hand, and the observer had to lean way over the side of the cockpit to take his photographs. Andrew J. Brookes, who chronicled the story of British air reconnaissance, reported:

> Many observers nearly fell out when they leaned overboard for the first time clutching this weighty contraption. . . . But even if the observer kept his safety belt on and his feet braced, his problems weren't over. By fumbling about in thick gloves or with frozen fingers he had to make eleven separate operations before he could expose the first plate, followed by ten more for each plate after that.[1]

By the end of that war vast improvements had been made; hundreds of photo planes with cameras mounted in fixed frames below the cockpits were in use, and these played significant roles in campaigns such as the Battle of Ypres and the final major offensives of General Erich Ludendorff in the spring of 1918.

During the intervening years there were still greater strides in photographic technology and equipment and by 1939 fast, long-range recon planes, with automatically triggered cameras built into nose and fuselage, had become a reality. But as the planes flew higher, faster, and deeper into enemy territory, bringing back great masses of visual information, the need quickly grew for interpreters who could read and decode the data locked away in the prints.

Aerial photographs speak a somewhat veiled language, and when the subtle nuances of enemy camouflage are added the language grows even more obscure. In World War II, understanding and deciphering this Niagara of raw visual material took a high degree of skill; and in the course of the conflict the Americans and British became far more adept at it than their counterparts in the Axis armies.

In the United States, Photo Intelligence (PI) training was centered at Lowry Field, Colorado and at the Air Intelligence School in Harrisburg, Pennsylvania. Officers trained at Harrisburg followed a strange curriculum, which to an outside observer may have seemed rather eccentric. The trainees learned among other things how to measure and analyze shadows and how to determine a ship's speed by the angle and appearance of its wake. They studied and memorized the tread patterns of tanks, the shapes of haystacks, the texture of foliage, the layouts of oil refineries, and the flow of freight through French railroad yards; and all this information fell surprisingly into place once the aerial photos started moving under the stereo viewers.

British photo interpreters were based at Medmenham near Henley, their work closely coordinated with RAF Headquarters at High Wycombe, west of London. American recon units, which first saw action in North Africa, came to England early in 1943 and were based at Mount Farm in Oxfordshire, also at High Wycombe and Benson Field. Here, under the direction of Colonel Elliott Roosevelt, they supplied target information for the strategic bombing missions of the U.S. Eighth Air Force. After D-day other photo units, commanded by Colonel James G. Hall, joined the Allied troops on the continent.

In Europe, close to the battle lines, PIs of the U.S. Ninth Air Force, the British Second Air Force, and special U.S. Army units, worked with the fighter-bombers in support of tanks and infantry. These small groups—each numbering thirty to forty men, and set up near the front lines in tents, barns, old shacks, and crumbling chateaus—operated around the clock, tracing troop and supply movements, analyzing bomb damage, and searching for hidden targets.

The cameras' eyes were relentless. Day after day, weather permitting, Allied pilots flew endless weaving patterns over enemy territory while automatic cameras in the bellies of their aircraft clicked away. The planes—chiefly American P-38s and British Spitfires—were stripped of armament and other superfluous weight, and depended for safety on speed and evasive action. Their cameras photographed miles of vertical overlapping strips, from heights varying between five thousand and twenty thousand feet, though at times the pilots flew "dicing" missions, skimming over enemy positions almost at ground level. When the planes landed, the exposed film was rushed to nearby field laboratories for processing into black-and-white prints, nine inches square, and these in turn were rushed, sometimes still wet, to the PI units. The overlapping of aerial prints made it possible to construct accurate photomosaic maps of huge sections of enemy territory. This process also meant that any *two* adjoining photographs could be placed together under a stereo viewer and studied three-dimensionally.

The Allies' photographic ouput was prodigious; the Ninth Tactical Air Force alone, flying constant recon missions from D-day to the final German surrender, produced some 13 million prints[2]—enough, according to one enterprising military statistician, to reach when placed side by side (should anyone so desire) from Chicago to San Francisco. The pilots who flew these sorties showed particular daring and enterprise. To gain speed and maneuverability the photo planes carried no weapons, and the fliers, periodically attacked by enemy fighters, had to rely on hair-trigger reflexes and fast evasive tactics. Many planes returned to their fields so badly damaged by flak and interceptor fire that they could barely make landings; but in every case, first thoughts were for the precious film cargoes the planes carried.

The RAF's recce units included aces such as Wing Commander G. W. Tuttle, Squadron Leader P. B. "Pat" Ogilvie, and Pilot Officer Adrian Warburton. Among the many notable American fliers were Lieutenant Colonel Russell Berg, Captain Charles Batson, Captain Robert Holbury, and the legendary Colonel Karl "Pop" Polifka, who later saw action as a brigadier general in Korea. Polifka, technically over age and not permitted to fly combat missions in World War II, often signed himself out on dangerous recon sorties under the name of "Lieutenant Jones"—which caused a slight problem when General Mark Clark decided to decorate the mysterious "Lieutenant Jones" for his excellent work in the Italian campaign. When Clark discovered the truth he reacted with good humor, enjoying Polifka's little scheme as well as the pilot's subsequent embarrassment.

Charlie Batson, another of the daredevils, flew a classic low-altitude photo mission over the Normandy coast shortly before D-day. Batson came in so low over the Wehrmacht fortifications that his wings brushed the trees; and when he returned in his battered P-38, air historian Glenn B. Infield noted:

> . . . there were, by count, 130 holes in the fuselage and wings, and bullets in the camera compartment and camera magazine; the oil radiators on both engines were stuffed full of twigs and leaves; the leading edge of the wing was damaged; the left engine was shot out and the supercharger on the other engine was lost.[3]

In January 1945, Bob Holbury flew a historic dicing mission for General Patton, whose Third Army urgently needed information about bridges and enemy defenses along the Saar River. The cloud ceiling was very low, but Holbury covered the entire sector almost skimming the water, and obtained over two hundred good photos for Patton's planned drive. When he landed, one of his two engines had been shot away and the entire plane was riddled with flak damage. According to Infield, "The maintenance crew took one look at the Lightning and pushed it into the junk pile."[4]

Perhaps the most prestigious of the war's photo pilots was the French author, Antoine de Saint-Exupéry whose unique, near-mystic devotion to flying was documented by him in several notable books.[5] Saint-Exupéry served as a reconnaissance pilot in the French Air Force during 1939–40, and after the fall of France went to Algiers and later to America. Returning to Tunisia in 1943, he volunteered for duty with a U.S. Air Force photo unit based at La Marsa. Flying had by then become the veteran aviator's overwhelming passion; though considerably over age, and unfamiliar with the complexities of piloting a P-38 Lightning, he pleaded and persisted, and was finally allowed to make a limited number of sorties. "Saint-Ex," as he was affectionately known to the Americans, lived in fear of being grounded. For months he continued his flights over southern Europe, pulling endless wires to keep himself on active service. Then in July 1944, he took off on a lengthy photo mission over Annecy and Grenoble, and his plane failed to return.

The work of airmen such as Saint-Exupéry, Warburton, Holbury, Polifka, Batson, and many others literally provided Allied field commanders with an extra dimension for planning their strategy and tactics. The visual cover they obtained—often at high risk—was in turn rapidly converted by the interpreters into reports classed as "first," "second," or "third phase," depending on their urgency and the amount of information involved.

Some but not all of this daily activity dealt with enemy camouflage: The bulk of the field reports covered such matters as the movement of troops and supplies, enemy airfield activities, the location of artillery, and the degree of damage caused by Allied bombers. In any bombing raid, the true results could only be assessed *after* the explosions and smoke had settled and the air had cleared, and for this good photos and good interpreters were needed. The Germans were also geniuses at making hasty patchwork repairs, and key Allied targets—even after heavy attack—had to be constantly monitored.

Is the bridge at Pontoise out of action or has it been repaired? What percent of the chemical plant at Bitburg has been destroyed; how badly is production crippled? Are all through-lines cut in the Kreuztal rail yard? What is the present condition of the bridge at Pontabault? At Bad Munster? At Colbe? At Volkmarsen?

Analysis of bomb damage comprised the chief work of the field detachments; but periodically, during their photo searches, the PIs would come upon suspicious-looking areas and then a unique camouflage decoding process would begin. Shadow formations, variations in foliage, tunnel openings, hidden gun positions, track patterns, access roads, and rail spurs—these were the valued clues. Bit by bit the jigsaw pieces of fact and surmise were fitted together, and if a reliable picture emerged—if the photographs "spoke" with enough clarity—a T (for Target) Report would be forwarded to the Tactical Air Command. Over 1,700 T Reports[6] were issued by American units during the war, covering field installations such as command posts, ammunition dumps, gun emplacements, supply depots, radar stations, and troop centers; and these led to scores of successful attacks by Allied planes and artillery.

German camouflage in the immediate combat zone varied in quality. Some attempts were hastily contrived and easily detected; others such as the "deserted" French barn complex were more ingeniously prepared; and some targets deep inside Germany were so well concealed that they remained hidden until discovered by Allied occupation troops after VE-day.

In the early *blitzkrieg* phase of the war the Wehrmacht generals had little interest in concealment and deception.

Blitzkrieg leaves no room for subtlety; with their powerful panzer divisions, motorized infantry, heavy field guns, efficient radio network, and lethal dive bombers the Germans simply crushed all resistance and rolled ahead. Then came Dunkirk, the Battle of Britain, and the eventual frustration of the Luftwaffe by the RAF. In the months that followed, unable despite their best efforts to win air supremacy in the west, the Nazis—on this front if not elsewhere—gradually found themselves on the defensive.

By 1941 the British had recovered and rearmed sufficiently to begin carrying out night air raids on factories and synthetic oil plants in the Ruhr valley. Night bombing at that point was still experimental, and the missions weren't particularly successful. "In those days," an RAF bombardier recalled, "it was a hit-or-miss situation, frequently miss." Harassed British pilots, guiding their Hampdens and Wellingtons through black skies lashed with flak and tracer bullets, often failed to hit their assigned targets and many never found their targets at all; but the British persisted and the night missions slowly grew in effectiveness.

In the summer of that year a photo expert at Medmenham, Lieutenant Douglas Kendall, noted something odd while screening new coverage of northwest Germany.[7] The photos, taken near the Ruhr town of Soest, showed three large rectangular structures arranged symmetrically in an open field. The walls of these curious oblongs stood (he estimated by measuring the shadows) about five feet high, but the objects had no roofs and seemed to be filled with bales of straw. Scattered over the area, in and around the oblongs, were scores of bomb craters.

Kendall carried the enigmatic photos to the desk of another interpreter, Lieutenant Geoffrey Dimbleby. "What do you make of these?" he asked.

Dimbleby studied the prints under his stereo viewer, then checked a large-scale map of the area and the latest strike reports from Bomber Command. Not far from the Soest rectangles was a German assembly plant, which had recently been raided. Within a few minutes the officers had found an answer.

"Fire sites, of course," Kendall nodded with satisfaction. "I've been on the lookout for *something* like this for quite a while."

The two interpreters were familiar enough with that particular ruse, since the British had used hundreds of Starfish decoys themselves, earlier in the war. The German version of the hoax was elementary but effective: As the defenders knew, RAF bombers invariably came over the Ruhr in waves—and when the first bombs started falling on or near a target factory, crews quickly set fire to nearby decoy "buildings" filled with combustibles. At night, from fifteen or twenty thousand feet up, the burning structures looked very much like a plant under heavy attack; and successive waves of planes, following the first, would be lured toward the blaze to drop their bombs far from the real objective.

At first the Nazi decoys worked extremely well. Pilots would return from their missions with glowing reports of great success; but on the following day the post-strike aerial photographs would show that the assigned target, now supposedly in smoking ruins, was fully intact and as productive as ever. This happened not once but repeatedly, until the mystery was solved by the PI reports from Medmenham; but the problem itself remained, and as the night bombings intensified the Germans came to rely more and more on defense by misdirection.

The Ruhr decoy system grew so widespread and elaborate—with complicated "plant layouts" and canvas "factory roofs" to fool the photo interpreters—that a special RAF Decoy Section under Lieutenant Dimbleby was finally organized. Dimbleby's unit concentrated not on the targets but on the areas that surrounded them. In daytime the fire decoys, for all their artfulness, weren't hard to locate: By their very nature they had to be placed in open, highly visible areas, and in proximity to legitimate war plants, most of which had already been plotted on RAF maps. Furthermore, the decoys never acquired the lively, hectic ambience of real operational factories, with their endless streams of trucks, busy road junctions, and the inevitable clutter of goods and supplies in adjoining storage yards.

As each new fire site in German territory was found, photographs and map coordinates were rushed to Bomber Command and the squadron leaders were carefully briefed. This helped to counteract the Nazis' efforts, though their deceptions continued to create problems for bomber pilots coming over at high altitudes with nothing but moonlight (and sometimes not much of that) to provide needed visibility.

During this same period, with the RAF beginning to show its growing fangs, the Germans also created a giant illusion in the city of Hamburg, which for size, scope, and drama rivaled anything ever attempted by the British or the Americans.

The Alster Basin in Hamburg is a large body of water situated at what was then the heart of the city's business and industrial life. The basin was traversed near its lower end by the Lombard bridge, which carried the main rail lines linking Berlin with key northern cities such as Bremerhaven, Lübeck, Neumünster, and Kiel. The old bridge acted as a divider between the outer basin, or Auber Alster, and the inner section, or Binnen Alster. Moonlight reflecting on this large inner, almost square-shaped body of water made an ideal orientation point for attacking planes. In addition, just a few hundred yards away was the Hamburg railroad station with its complex lines feeding directly onto the Lombard bridge.

This combination of bridge and rail station was an irresistible, clearly marked target and the Germans were anxious to protect it. So to divert the night raiders—and throw the RAF bombardiers off the track—they launched a plan to conceal the entire Binnen Alster, comprising some *225,000 square yards of open water.* Hundreds of poles were driven into the relatively soft bed of the basin as supports for acres of wooden sheeting. The top of this wooden canopy was then painted to simulate the rooftops of dwellings and office buildings, as well as the streets of Hamburg.[8]

Farther north, to complete the illusion, a fake bridge more than half a mile long was built across the outer Alster, exactly simulating the Lombard bridge. It too was built of wood, complete with imitation railroad tracks and a dummy "train" made of fiberboard. This in effect shifted the entire Binnen Alster hundreds of yards to the north, which would, of course, throw off bombers using it as an aiming point for the bridge, the railroad station, or the main industrial area.

The vast camouflage project took four months to build, from January to April 1941, and involved hundreds of technicians and carpenters. It proved convincing on aerial photographs, but British Intelligence, alerted both by secret agents and their own efficient PIs at Medmenham, were well aware of it. However, the ruse still served its purpose: It denied RAF fliers easy orientation and caused a few vital seconds of delay and confusion, which was made still worse by heavy German defensive fire.

This "relocating" of the huge Binnen Alster was a classic and impressive use of misdirection, particularly for night raiding missions, but its value was limited. In December 1941, the entrance of the United States into the war brought a tremendous change and expansion in Allied bombing operations. American factories tooled up quickly and were soon turning out hundreds of powerful long-range and medium bombers. The British by then had begun using their new four-engined Avro Lancasters; and during 1942 these were gradually joined by large formations of B-24 Liberators and B-17 Flying Fortresses.

By 1943, the Allies were carrying out massive 1,000-plane strategic raids on military targets in all parts of the Third Reich. The American planes usually bombing by day, and the British by night, poured countless tons of explosives on German rail yards, airfields, coast defenses, war factories, supply depots, and major industrial cities such as Hamburg, Dortmund, Essen, Frankfurt, Bremerhaven, Cologne and Berlin itself, which the boastful Goering had once insisted would never feel the effects of a single Allied bomb. In Berlin, partly for the sake of civilian morale, an attempt was made to hide key aiming points such as the Kaiserdamm Exposition Halls and sections of the Tiergarten and Unter den Linden under huge camouflage nets. But the Allies with their giant formations of planes had begun using saturation or "carpet" bombing patterns, blanketing enormous areas and making camouflage efforts generally superfluous.

Hamburg, where so much time and energy had gone into transforming the Alster Basin, now became the recipient along with other cities of widespread devastation. But for the Allies the air offensive had a grimly realistic purpose: It had been launched, Eisenhower noted in his memoirs, because ". . . we were convinced that this policy would result in shortening the war and therefore in the saving of thousands of Allied lives."[9]

The mass raids went on in strength; then in the spring of 1944 Albert Speer, Germany's minister for war production, was faced with still another critical problem. In a last-ditch effort to regain the initiative, the Nazis had been racing to develop jet fighter aircraft—one of a long series of "miracle weapons" on which Hitler's fading hopes were pinned. Their deadly Me-163 and Me-262 jets could indeed outclimb, outfly, and outfight anything the Allies had in the air at that time. The problem was, how could these crucial new planes be safely produced? Speer and his staff were all too aware, painfully so, of the fate of German factories; any conventional plant that attempted to manufacture jet aircraft would soon be bombed into rubble.

After much discussion and debate an answer was found: The minister, with his genius for organization, arranged for production to be parceled out to small, innocuous, widely dispersed facilities, using natural concealment wherever possible. Speer's resourceful managers made use of everything: old school buildings, private sanitariums, tanneries, wine cellars, garages, cigar and candy factories, underground caverns, and unfinished autobahn tunnels. In all, according to Andrew Brookes, the components of thirty-two large aircraft factories were scattered over more than *720 separate locations.*[10] So in this final phase of the most technologically advanced war in history, German jet aircraft production became a cottage industry.

Firebreaks were also used. Years earlier, great swathes had been cut through the thick pine forests of Bavaria and Thuringia as a method of controlling the spread of forest fires, and they fitted in perfectly with the German plan. Camouflage netting was stretched over these open lanes, then covered with coniferous foliage taken from the surrounding woods. From the air and on photographs, this exactly matched the real growth in texture, color, and density. When the netting was extended to the ends of a particular firebreak, it was almost impossible to detect from above, and under this cover a makeshift factory using slave laborers could be set up.

The Allies learned of these activities from secret agents, war prisoners, and liberated foreign workers and the recce units were soon called on to find the hidden plants. Faced with this great challenge, the PIs at Medmenham and Mount Farm fell back on a classic technique: the study of track prints.

Every viable factory depends on its access roads, much as a heart depends on a network of veins and arteries. Somehow or other, supplies have to be brought into the plant and finished parts have to be shipped out. Of course, all this transport could be handled at night, and was, but it remained an Achilles heel in the Speer operation. When the interpreters found an area that seemed even vaguely suspicious, they searched for telltale loops on the ground indicating truck turnarounds, or the sharp double "V" patterns left by wheels coming to a stop, then backing away again. Open fields near "abandoned" Luftwaffe bases were closely watched. Flight Officer Constance

Babington-Smith, the RAF's noted photo interpreter and expert on German aircraft, described this kind of track activity as,

> . . . the accumulation of faint lines that tells an interpreter where feet or wheels have been passing over the field again and again—faint pale lines which are actually caused by the reflection of light from the myriad flattened blades of grass.[11]

Visual images of this kind were an important signal. Ordinary dirt roads were also carefully studied. If one or more roads seemed to emerge for no apparent reason from an innocent patch of forest, they were traced for miles across the countryside to see where they went. Such roads, if they were being used by vehicles carrying aircraft parts, invariably led to railroad warehouses, highway depots, or river transshipping points. Sometimes a poorly hidden cache of factory supplies would be found, setting off another chain of deductions. The PIs also used comparative cover, matching earlier photos with current ones of the same area. The slightest unexplained change in appearance—the smallest variation in terrain detail—could provide a meaningful clue; and little by little, using all these resources, the interpreters pried the data from the photographs.

Many of Speer's well-hidden mini-factories escaped detection; but numerous others were found by the PIs, reported, and destroyed. Despite all this, the Germans still succeeded in turning out about one thousand jet planes by the war's end; yet only a small number ever went into action. Because of the heavy bombing of synthetic oil plants, there was an acute shortage of fuel for the Me-163s and 262s. Moreover, the longer runways needed by jet aircraft—and their telltale scorch marks—could easily be spotted from the air; and fighter-bombers, aided by T Reports, destroyed hundreds of the dangerous new planes before they ever left the ground.

Despite the inevitable problems and frustrations, Allied photo detectives pitted against the tenacious German conjurors achieved a brilliant degree of success. In all the combat zones the Americans and British, with total control of the air, had eyes everywhere. Recon planes of General Hoyt Vandenberg's Ninth Air Force, for example, flew thirty-one thousand separate sorties during the war,[12] resulting in a mass flow of prints that helped interpreters probe everything that occurred behind enemy lines, and won high praise from Allied commanders in Europe. Discussing this factor in the campaign, Eisenhower wrote, "Airplane photography searched out even minute details of defensive and offensive organization and . . . information so derived was available to our troops within a matter of hours."[13]

In certain instances, such as the hunt for Speer's hidden jet plants and the baffling search for Germany's secret V-weapon sites, the PIs ran into many obstacles, but managed to overcome most of them. The great flow of aerial photographs consistently spoke with eloquence—and when they did, the Allies listened.

10

The Defense of Target 42

1943–1944

The London morning was deceptively peaceful. General Sir Stewart Menzies, chief of MI-6—the agency which had been wrestling with the problem of Germany's secret weapons—left his office shortly after 11:00 A.M. and started toward the Royal Military Chapel nearby. Menzies, on his way to attend the service, was late; inside the chapel the congregation had just risen to its feet to sing the *Te Deum*. The handsome nineteenth-century building, known also as the Guards Chapel, stood a few hundred yards from Buckingham Palace, Parliament, and 10 Downing Street, and that particular Sunday it was crowded with worshipers.

As Menzies passed St. James's Park and hurried across Birdcage Walk, he heard a harsh droning sound overhead. The grating noise grew louder and louder, then stopped abruptly. The general glanced anxiously toward the sky and dove quickly for cover. Seconds later, forty-five hundred pounds of death tore through the roof of the little chapel and exploded in a chaos of smoke and flame. Menzies's tardiness possibly saved his life, but 121 people—mostly soldiers and their families—were killed by the bomb, many others were seriously injured, and the building was almost totally demolished.[1]

This sudden attack took place on June 18, 1944, twelve days after the launching by the Allies of Operation Neptune. Other such bombs had already fallen on London and its suburbs, but the tragedy at the Guards Chapel dramatized a fearful new menace that had come to plague the people of England.

Hitler's *Vergeltungswaffen,* or vengeance weapons, were known as the V-1 flying bomb or "buzz bomb," and the V-2 rocket. The V-1, a pilotless ramjet missile, carried an 1,870-pound warhead and had a range of about 160 miles. With its stubby sixteen-foot wing-span, it looked much like a miniature airplane flying without benefit of human guidance. The V-2, a liquid fueled rocket, carried 2,150 pounds of explosives and had a maximum range of 210 miles. In flight the buzz bomb had an odd characteristic drone, which Corporal Gerald Bloch of the U.S. Ninth Army Combat Engineers compared aptly to the sound of ". . . an almost empty coal truck rumbling over a cobblestone street." This harsh noise suddenly stopped as the bomb began to fall, which gave a few precious moments of warning; but the sinister V-2 gave no warning at all. Launched from an upright position, the rocket raced to a trajectory of 50 or 60 miles, then plunged silently toward its target with incredible speed and power.[2]

Together, these deadly and at that time unique weapons represented the German dictator's last significant hope for victory in the war. In June 1943, he had told his military leaders that ". . . the Germans have only to hold out." With the coming of the vengeance weapons, Hitler informed them, ". . . London will be leveled to the ground, and Britain forced to capitulate."[3] Of course nothing of the sort happened or came close to happening; but the V weapons added a brutal dimension to the war, caused aimless death and destruction, and strained the morale of the weary British people almost to the breaking point.

The German code designation for London was "Target 42," a target which had long been one of Hitler's major obsessions. His V-1s first fell on London a few days after D-day; and within two weeks, Cave Brown reported:

> . . . the Germans had launched more than 2,000 bombs, hundreds of which penetrated British defenses and fell upon London and the surrounding areas. Destruction and dislocation were severe, and the government began the evacuation of women, children, the elderly, and hospital patients that would total nearly one million people.

The evacuation, he added, "placed an enormous strain on roads and railways that were already carrying capacity loads to the ports for the battle of Normandy."[4] SHAEF, heavily burdened by the incredible demands of supplying and protecting the vulnerable beachhead in France, did all it could to defend Target 42, and Eisenhower quickly decreed that attacks on the V-1 sites ". . . would now take first priority over everything except the urgent requirements of the [Normandy] battle."[5]

The campaign to find and destroy these missile bases, known as "Operation Crossbow," had actually gone on for many months, and no other phase of the conflict so tested the abilities of British-American intelligence plus the combined striking power of the RAF and the U.S. Eighth and Ninth Air Forces. Pitted against this great array of strength were endless Nazi reserves of energy, tenacity, and resilience that transformed Crossbow into a frustrating cat-and-mouse game filled with accident and miscalculation, error and insight, paradox and uncertainty, good luck and bad—plus brilliant camouflage by the Germans and equally brilliant detection work by the photo sleuths at High Wycombe, Mount Farm, and Medmenham.

The Allies had first learned, in a general way, of the existence of Germany's terror weapons as early as November 1939. At that time a mysterious package was found lying in the snow on the grounds of the British Embassy at Oslo. A cryptic note inside suggested that the information could be of use to the British authorities, and was signed simply, "a well-wishing German scientist."[6] The identity of the well-wisher was not—and never has been—discovered, but the package was rushed by diplomatic pouch to MI-6 Headquarters in London. Its contents, which became known as the "Oslo Report," gave considerable information about German radar experiments, new acoustic torpedoes and most important of all, an unusual rocket program. The Wehrmacht, said the Oslo Report, were already testing long-range rockets on the island of Usedom in the Baltic Sea, just off the coast of Germany. The missile testing site was called Peenemünde.

Dr. R. V. Jones of MI-6's scientific department, an expert on German weaponry, analyzed the Oslo Report in detail; he was impressed with it and felt it should be explored further, but other strategists remained skeptical. The idea of guided ballistic missiles seemed, in 1939, to belong more to the world of science fiction than to science. Of course rocket tests and experiments had been conducted since the 1920s—particularly by the American inventor, Dr. Robert H. Goddard—but this work was rudimentary and its application to military weaponry was (outside the Third Reich) a technical feat almost too complex to grasp. Besides, the Germans—like the English—were clever at hoaxes. Wasn't it possible that the Oslo Report was a deliberate fraud? A counterfeit planted to fool the British and lure them into a wasteful, time-consuming wild goose chase? Some of the analysts agreed with Dr. Jones while other supported the "hoax theory"—but soon more pressing problems intervened.

In the spring of 1940 the Germans attacked western Europe, then came the British evacuation at Dunkirk, and after that the desperate air war between Goering's Luftwaffe and the RAF. At MI-6, deluged with daily crises and problems, the Oslo Report was quietly filed away and largely forgotten. But as the war continued, new and disturbing reports kept trickling in. British undercover agents sent word from Berne, from Lisbon, from Istanbul. Other messages came from resistance units in the occupied countries. These told of mysterious "construction projects," "launch sites," and ominous but unidentified "earthworks" being built along the Channel coast, many of them in the Pas de Calais. Remote-controlled missiles—the fantasy weapons of science fiction—were soaring at last into the real world, and early in 1943 General Sir Hastings Ismay, Churchill's military secretary, sent the prime minister a memo which confided, "The Chiefs of Staff feel that you should be made aware of reports of German experiments with long-range rockets. The fact that five reports have been received since the end of 1942 indicates a foundation of fact even if details are inaccurate."[7]

With Churchill's approval, Duncan Sandys of the Ministry of Supply was put in charge of a new investigation, and the photographic hunt for Hitler's secret weapons went into high gear. But what were the Spitfire recce pilots supposed to photograph? And what were the hardworking PIs at Medmenham supposed to look for? At that point, nobody was quite sure. Their fragmentary (and often contradictory) information had led Menzies's scientists to expect giant forty- or fifty-ton rockets requiring huge installations served by vast networks of spur lines and rail heads, but none of these rail lines were in evidence. A further complication came from the fact that the Nazis were simultaneously developing *three* distinct types of *Vergeltungswaffen*. One was the V-1 flying bomb, another was the V-2 long-range rocket, and the third was a gigantic kind of cannon called the "London gun," also known as the V-3.

The London gun was a nightmare weapon with a barrel over four hundred feet long. Charges were designed to explode at intervals inside this huge barrel, driving heavy shells across distances up to one hundred miles. The giant guns, buried underground with only the ends of their muzzles showing, would be protected by concrete roofing eighteen feet thick;

and from such strongpoints they would automatically pour death on London at the incredible rate of one shell every ten or twelve seconds. The first of these extraordinary batteries, being installed in a small French town behind Calais called Mimoyecques, involved over five thousand German engineers and slave laborers. The guns were never completed, but the mysterious activity at Mimoyecques, carefully photographed and studied, only added to the interpreters' confusion.

Later, when Crossbow was fully under way, U.S. Air Force units played an important role; but since the British were responsible for the air defense of England, initial investigations were carried out by the RAF. Constance Babington-Smith, the WAAF Flight Officer who was the first person to spot a V-1 bomb on an aerial photograph, described those difficult early days:

> Both in London and at Medmenham it was a time of frustrating confusion in the secret weapon investigation . . . a time of groping in the dark, of trying to lay foundations in a swamp. It was as though the parts of two or three jigsaw puzzles had been jumbled together, and it was sorely tempting to try to find only *one* answer, only *one* weapon. It seemed a triumph when two or three bits of puzzle fitted together and could be identified as "Rocket" —and it was all too easy to ignore the bits which did not fit in with these.[8]

But despite their doubts and conflicting viewpoints the British kept trying. Peenemünde, that mysterious enclave on the Baltic coast, was photographed repeatedly; so too were French installations at Watten, Sottevast, Siracourt, Bonnetot, Abbeville, Wizernes, Nucourt, Lottinghem—and of course the puzzling construction at Mimoyecques. At that time German camouflage was fairly primitive and the British PIs had no trouble following the Nazis' activities.

Aerial cover showed that the work at Peenemünde was

expanding; in July the interpreters reported that antiaircraft defenses at the experimental base had been strengthened and smoke generators had been installed, which could be used to conceal the target during a bombing raid. In addition the Germans had created a large, elaborate decoy site nearby, covering some twenty acres. Up to that point there were still a few analysts who felt that the Peenemünde "installation" was merely a scare plan, part of an insidious war of nerves against the British. But the discovery of the decoy backfired on the Germans and brought unintended results: It convinced the doubters that the Nazis indeed had *something* of great value at Peenemünde, something which they wanted to protect.

The growing threat obviously had to be dealt with, and on the night of August 17, 1943, RAF Bomber Command launched a massive low-level raid on Peenemünde in which six hundred bombers dropped over eighteen hundred tons of fire bombs and high explosives. The British lost forty planes on that mission, but the damage to Peenemünde was enormous and the raid has been described as one of the most decisive air operations of the war. Because of it, the paranoid dictator's plans for "revenge" against England were set back seriously—according to some estimates for as long as six months. Ten days later the U.S. Eighth Air Force bombed the big installation at Watten (which was later found to be a V-2 storage and launching shelter), and here too the damage was so severe that the site was abandoned by the Germans.

Meanwhile the search continued; the PI units had been ordered to "watch northern France," and with the help of their camera planes they methodically studied mile after mile of French countryside. "It was harder than searching for a needle in a haystack," one interpreter remarked, "because we at least knew what needles and haystacks looked like."

Hundreds of sorties involving thousands of aerial photographs were flown over suspected regions, and new evidence gradually appeared. Some of the photos showed earthen ramps under construction, most of which appeared to be aligned on London. One site, almost completed, was near the town of Yvrench in a wooded area called the *Bois Carré*. Here—along with the mysterious ramps—were several buildings with a peculiar curve at one end. To the PIs the buildings resembled giant skis lying on their sides, but what was the purpose of the odd structures? Although nobody had a decisive answer the recon flights went on, and more *Bois Carré*-type "ski sites" were found and carefully tagged on Bomber Command's target maps.

Then Flight Officer Babington-Smith, combing through some earlier coverage of Peenemünde, chanced upon the link that tied many of the baffling strands together: There in the testing area she noticed a tiny blurred object that suddenly took on great significance. "The quality of the photographs was poor," she wrote, "but even with the naked eye I could see that on the ramp was something that had not been there before. A tiny cruciform shape, set exactly on the lower end of the inclined rails—a midget aircraft actually in position for launching."[9]

A midget aircraft actually in position for launching. This was the needed breakthrough, the first visual proof, and it galvanized the hunters. By careful measurement and analysis they deduced that the strange ski buildings were actually storage and assembly points for the kind of flying bomb that Babington-Smith had just discovered. All the evidence pointed to it, and the confusing jigsaw pieces began to make sense.

By now over ninety such ski sites had been mapped, and many of them seemed almost ready for operation; so in December of that year, with the Normandy landings only six months away, the Allies struck at the Calais weapon bases with every plane they could spare. Staff Sergeant Alvin Katz, crew member of an Eighth Air Force Heavy Bombardment group stationed near Kettering, recalled the feelings of the airmen during those weeks. "Our main mission was to pave the way for the invasion by hitting industrial plants, rail yards, and aircraft factories," he said, "but the secret weapons were on everyone's mind. Flak and enemy fighters were problems we could deal with, but rockets and flying bombs were like robots—blind killers. They were diabolic, and we came to hate them."[10]

Under command of Lieutenant General Carl Spaatz and Air Chief Marshals Arthur Harris and Trafford Leigh-Mallory, the Crossbow attacks went on day and night. There were heavy losses to the Allies, and one of these casualties was Lieutenant Joseph Kennedy, a brother of John F. Kennedy. On a special mission to bomb the mysterious Mimoyecques base Lieutenant Kennedy's B-24 Liberator, packed with explosives, blew up in mid-air and both he and his copilot were killed.

Scores of other fliers and aircraft were also lost, but the attacks kept up relentlessly. Between December 1943 and D-day, Allied planes dropped over thirty thousand tons of bombs on numerous sites, supply points, and manufacturing facilities, and historian Basil Collier noted that this was ". . . more than half as large again as the whole weight of bombs . . . which the Germans had aimed at London in major attacks during the eight months of the 'blitz' from September 1940 to May 1941."[11]

The first phase of Crossbow brought quick results: Within a month the entire German system of launching bases in the Pas de Calais was a shambles. "By the beginning of 1944," noted Andrew Brookes, "every *Bois Carré* site apparently nearing completion had been destroyed, and the Germans, recognizing the conspicuity of their ski site layout, decided to abandon the rest."[12]

At SHAEF and the British War Office there was great relief at this apparent end of a dangerous threat, but the mortal cat-and-mouse game was far from over. The Nazis reeled under the bombings, then rallied with energy and shrewdness. Wehrmacht General Erich Heinemann, chief of the weapons program, pulled his *Flakregiments* out of the wrecked sites and began building new ones, and this time the Germans were wary. They simplified the bases drastically, elaborate structures such as the ski buildings were eliminated, and only the most essential launching equipment retained. All of the new sites were carefully, subtly camouflaged; most were dispersed deep in wooded areas between Le Havre and Calais, supplies were stored underground, ramps and control points were hidden with nets and foliage, and all vehicles were kept under cover. The Germans used only existing roads, and equipment was moved late at night or in bad weather when there was no likelihood of Lightnings or Spitfires appearing overhead.

Wehrmacht security was tightened, leave for the *Flakregiments* was restricted, and letters home were heavily censored. The obsession with secrecy was so great that a certain Colonel Wachtel, the unit field commander, took to using a fancy pseudonym and went about on his inspection tours wearing a ludicrous false moustache. His disguise actually fooled no one but presumably gave Wachtel some degree of solace.

The Nazis also made use of decoys: French laborers were put to work repairing some of the original sites as well as a number of old supply areas. General Heinemann had no intention of using any of these, but hoped to mislead resistance agents who would surely report such activities to the Allies.

All this information soon found its way to MI-6, and the photo interpreters—now augmented by American units—went wearily back to work. The camouflaged sites, scattered and well hidden, were extremely hard to locate; but extra sorties were flown and the PIs, helped by intercepted radar signals and tips from the underground, scanned the new coverage print by endless print. A faint track, an odd shrubbery arrangement, a few unexplained supply crates—all these became useful clues to be analyzed, discussed, investigated.

The search was dogged and slow, and it wasn't until April—with D-day a mere two months away—that the first of the heavily camouflaged sites was definitely identified. During the next two weeks another nineteen V-1 bases were found and once again the Allies felt the pendulum swinging in their favor. But now Heinemann added a subtle twist to his concealment tactics. Postwar evidence indicates that, under cover of darkness, a number of the old sites—ones which had been totally abandoned and ignored—were secretly reoccupied, and all the original damage was left precisely in place. The launching crews simply worked around and under the debris, without moving a single charred plank, twisted girder, or piece of broken concrete. This unique deception, which might be termed a "reverse decoy," was almost impossible for the PIs to detect through their stereo viewers or by comparative study of their aerial prints. In addition, bombs and ramps for these and the newer sites were transported in prefabricated modules, ready to be assembled and used at the very last moment.

In spite of all these ingenious maneuvers the Allied intelligence net succeeded, by the beginning of June, in locating over *sixty* of the hidden bases. It was a remarkable achievement, but by that time every U.S. and British plane and bomb was needed for the huge cross-Channel assault and the Fortitude deceptions which supported it, and of necessity the weapon sites were given a low priority. A few V bases were hit during Neptune, but a large number escaped destruction and shortly after the D-day landings, Heinemann's crews began their fatal attacks on Target 42.

Meantime, while the Allies concerned themselves with V-1 sites in the Pas de Calais, production was being rushed on the German V-2 rockets. After the great raid on Peenemünde, Hitler ordered the transfer of all important rocket research to the town of Blizna in southeast Poland, beyond the reach (at that point) of Allied planes. Actual manufacture of the missiles took place in two vast tunnels dug into the Harz Mountains. The tunnels, each more than a mile long and linked by dozens of connecting galleries, employed some eighteen thousand workers, an arrangement which has been described as the largest underground factory in the world.

By a twist of historic irony, the guiding genius of the German rocket program was a young scientist named Wernher von Braun.[13] Serving under Major General Walter Dornberger, military director of rocket operations, von Braun was considered absolutely indispensable to the success of the program. At one point the scientist resisted Heinrich Himmler's attempts to bring Peenemünde under Gestapo control, and as a result was arrested and thrown into prison for several weeks. It was only after a highly agitated General Dornberger convinced Hitler that the V-2 project would collapse without him that he was finally released.

Von Braun, Dornberger, and a hundred other key engineers surrendered to the American Army early in 1945, and were flown to the United States where they soon resumed their technical research. Eventually the rocket expert became a director of space flight at NASA, and was largely responsible for the brilliant development of the Saturn 5 rocket and the highly successful Apollo space missions. A pure scientist with no military or political alliances, von Braun tried to detach himself from Hitlerian ideology and to concentrate on space technology and research for its own sake; nevertheless his willing contribution to the Nazi weapons program was prodigious, and he bears major responsibility for the perfection of the deadly rockets.

In its early stages, due to trial-and-error setbacks plus the Allied bombing of supply depots, V-2 research lagged considerably behind schedule, and it wasn't until September 1944—months after the Normandy landings—that the first rockets were ready for use.

While most of the V-1 sites had been concentrated in the Pas de Calais, rocket operations were centered in Belgium and Holland, chiefly in the Dutch coastal city of The Hague. The photo interpreters, hard pressed, struggling to uncover the camouflaged V-1 bases, found the challenge of the V-2s even more difficult. For one thing, the Allies had assumed that heavy ballistic missiles, powered by ten tons of liquid fuel and carrying a one-ton warhead, would have to be launched from huge gantries served by special rail systems. In actuality, the Germans had designed scores of mammoth tractor-trailers called *Meillerwagen,* each of which could carry a complete rocket *plus all the components needed for launching.* There was no need for a permanent and highly visible launching base, or *any* prepared firing area. As Dornberger pointed out in his memoirs, the rockets could be launched ". . . from a bit of planking on a forest track, or the overgrown track itself."[14] There was also no need—as with the buzz bomb sites—for careful camouflage; the wooded areas in The Hague, and the many trails running through them, provided ample concealment for the mobile, self-contained *Meillerwagen.* After a rocket was fired, crew and tractor simply melted back into the accommodating protection of the tall, thick trees.

"Unless," wrote Basil Collier, "a rocket was being placed in position or fueled and serviced, there would be nothing to see and nothing to attack, except that troops and vehicles might be encountered more or less fortuitously."[15]

Constance Babington-Smith also recalled the helplessness everyone felt at Medmenham:

> The only hope of spotting a V-2 launching site was if a photographic aircraft happened to pass overhead when a rocket was set upright being fueled. Otherwise there was nothing to see. . . . It was a one in a million chance, and even if it occurred it would not provide a target for attack, as the "sites" were completely mobile.[16]

The resourceful Dornberger used his rocket crews as though they were guerrilla units—striking, evading, and withdrawing. At one point a cluster of *Meillerwagen* were reported—and photographed—in a large park in The Hague known as the Haagsche Bosch. The park was quickly and heavily bombed, and the rocket units were just as quickly withdrawn to an adjacent forest near Duindigt.

Reacting to this elusive game, the Allies decided to concentrate their attacks on underground caverns that served as rocket supply bases. "They were quite hard to hit," a Lancaster pilot acknowledged, "but at least they stayed in one place." Following Eisenhower's directive of June 18, 1944, Allied heavy and medium bombers, as well as fighter planes, hammered the rocket stores repeatedly. The target boards showed recurrent names: Raaphorst . . . Walcheren . . . Ter Horst . . . Eikenhorst . . . Staveren . . . St. Leu d'Esserent. Underground caches in the valley of the Oise were also hit, and again the unfinished installation at Mimoyecques. All these attacks further undercut the weapons program and reduced its effectiveness by an estimated two-thirds of capacity. But the remaining one-third, the deadly survivors, caused havoc and destruction.

The first V-1 bombs reached London at 4:00 A.M. on June 13, 1944; they kept coming for months afterward, and nearby areas such as Croydon, Greenwich, Lambeth, and Purley were also hit. Under direction of Air Marshal R. M. Hill, in charge of Home Air Defense, the British reacted vigorously. The protection of Target 42 was organized in three echelons. Fighter aircraft patrolled the English Channel, ready to intercept and shoot down the relatively slow-moving buzz bombs as they flew to the target. Sometimes a daring RAF pilot would even manage to fly close alongside a bomb, tip it up with his wing, and deflect it back toward Calais. Behind this fighter screen, along the southeast coast from Eastbourne to St. Margaret's Bay, was a vast belt three miles in depth containing eighteen hundred antiaircraft guns. And closer to London, sweeping from Limpsfield northeast to the town of Cobham, was a network of barrage balloons. The balloons, trailing heavy cables, were deployed so as to trigger or deflect flying bombs that penetrated the outer defense lines. Later, when the V-2s began coming over, the balloon barrier was extended.

Hill's tri-level defense strategy proved effective. In all, over eighty-five hundred V-1 bombs were launched against

England, about one thousand of which exploded in flight or fell into the Channel. Of the remainder, more than thirty-nine hundred were brought down by the British defenders. In his history of that battle Churchill noted, "With the new radar and predicting equipment and, above all, with the new proximity fuses, all of which we had asked for from America six months before, the performance of the gunners exceeded all our hopes. By the end of August [1944], not more than one bomb in seven got through to the London area. The 'record bag' was on August 28, when ninety-four bombs approached our coast and all but four were destroyed. The balloons caught two, the fighters twenty-three, and the guns sixty-five."[17]

Fewer rockets than buzz bombs reached England but the damage they did was greater; on the average, each rocket caused about twice as many casualties as a flying bomb. The first V-2s struck on September 8, 1944, about the time that British and Canadian divisions were overrunning the main V-1 sites in France. Of 1,360 V-2 rockets launched, about 500 hit London and an equal number fell on adjacent towns and communities. The rest were defective and failed to reach the coastline.

Hitler's bloody appetite for retaliation led to a total casualty list of some 8,900 British civilians killed and 24,500 seriously injured by the two secret weapons, and thousands of homes were destroyed.[18] In addition, almost 2,000 Allied airmen died in the battle to protect London. Belgium also suffered from the *Vergeltungswaffen*; as the Wehrmacht withdrew in frustration before the Allies' offensive, they launched great numbers of V-1s and V-2s against Antwerp, Brussels, and Liège, killing 3,400 Belgian civilians and hundreds of Allied servicemen.

The casualty figures, grim enough, would have been far greater if it hadn't been for the persistence of the photo detectives and bomber crews who, between them, managed to delay and contain—if not eliminate—this fantastic threat to England and the invasion plans. In his memoirs, Eisenhower speculated:

> If the German had succeeded in perfecting and using these new weapons six months earlier than he did, our invasion of Europe would have proved exceedingly difficult, perhaps impossible. I feel sure that if he had succeeded in using these weapons over a six month period, and particularly if he had made the Portsmouth-Southampton area one of his principal targets, Overlord [the liberation of northern Europe] might have been written off.[19]

The sector referred to by Eisenhower served as the chief staging area for Operation Neptune, and German attacks concentrated there could have done appalling damage to the invasion buildup; but by that time Hitler was blind to the tactical meaning of his V bombs, and could think of them only as a brutal new way to punish the British civilian population.

In deploying their weapon sites after the initial phase, the Germans made superb use of camouflage and dispersal, taxing the ingenuity of the Allied interpreters who had to explore, search out, and pinpoint these small, deadly, well-concealed bases, just as they would do with the jet plane mini-factories later in the campaign.

One strange repercussion of Crossbow—in a war replete with irony—deserves special mention. By a curious psychological process, the Allied failure to wipe out all the V bases actually helped *to insure the success of the Normandy landings.* Hitler had counted desperately on German terror weapons to crush England and turn the tide of the war, and their continued presence in the Pas de Calais fed these tenuous hopes. Speer, in the secret diaries he kept at Spandau prison, recalled a meeting between Hitler, himself, and several other high officials shortly before D-day. The bloodthirsty dictator was enraged at the Allied bombings of German fuel plants, and spoke of vengeance. "Soon we can begin the attacks on London with the V-1 and the V-2," he exulted. "A V-3 and a V-4 will follow, until London is one vast heap of ruins. The British will suffer. They'll find out what retaliation is!"[20]

As long as the V-1 sites existed the Allies, so Hitler believed, had no choice: they *had* to invade the Calais area and try to eliminate them. Every flying bomb exploding in London bolstered the führer's conviction that the Allies would attack the Pas de Calais; and this made him doubly susceptible to the Quicksilver and FUSAG hoaxes which implied the same thing. Thus the deception plans meshed with his own obsessions and strengthened his refusal to rush extra divisions to Normandy until it was—for his purposes—too late.

"I have a sixth sense about these things," Hitler was fond of announcing to his fawning generals. As the war intensified, wish supplanted reality, and like a V-2 out of control the dictator's fury doubled back on him, contributing to his eventual destruction.

11

Sorcery in North Africa

1940–1942

In the House of Commons, Churchill referred to him as "a daring and skillful opponent" and "great general."[1] His fellow officers in the Wehrmacht described him as a tough one hundred percent soldier and tactical genius. General C. J. Auchinleck, commanding Britain's Middle East Forces, complained in a top-secret memo that his opponent was being credited with "supernatural powers," and becoming "a kind of magician or bogey-man to our troops."[2] Anthony Cave Brown characterized him as "a war lover . . . narrow, completely dedicated and apolitical."[3] And Desmond Young, the German general's biographer, called him "a phenomenon; a nonpareil,"[4] and noted that even the British Tommies spoke not in anger but with a kind of soldierly admiration when they referred (as they often did) to the doings of "that bastard, Rommel."

Erwin Johannes Eugen Rommel was certainly neither a magician nor a supernatural force, but he did possess courage, daring, acutely single-minded professionalism, and a battlefield charisma similar to that of Allied generals such as Bernard Montgomery, Douglas MacArthur, and the singular George Patton. These attributes shaped the Wehrmacht leader's legendary quality, and the legend persisted from early 1941 when Rommel took over command of the Afrika Korps, through various military shifts of power, to his final historic defeat at El Alamein and stubborn retreat across the sands of Cyrenaica.

The battle for North Africa began as a confrontation between Britain's Army of the Nile and the Italian forces of Marshal Rodolfo Graziani, and ended with the remnants of the Afrika Korps trapped and crushed between the powerful pincer jaws of American and British divisions; and during the intervening months the fighting seesawed so erratically that Desmond Young likened the military results to "the recordings of a demented seismograph."[5]

Those dramatic desert campaigns, involving exotic locales such as Tripoli, Benghazi, Gazala, Tobruk, Sidi Barrâni and Mersa Matrûh, and culminating in the great clash at El Alamein, have been documented and analyzed in much detail, but little is generally known of the remarkable part played in the battles by camouflage and deception. All through the North African fighting—and particularly at Alamein—hoax and subterfuge were used more extensively by both sides than in any other theater of World War II combat; and in almost every case the ruses were used as *aggressive weapons:* decoys, feints, ambushes, fake strongpoints, false buildups coupled with camouflaged troop concentrations, were all neatly woven into the plans not only of the Allied commanders—Wavell, Auchinleck, and Montgomery—but also of Rommel the "Desert Fox," the master tactician.

The opening salvo of this little-known illusionists' war took place in September 1940, when Marshal Graziani's armies crossed the Egyptian border from Libya and attacked a much smaller British force under General Sir Archibald Wavell. The British, heavily outnumbered and outgunned, fell back along the coast toward Sidi Barrâni where they dug in to await reinforcements which were then being rushed from England. Wavell's problem was a difficult one: Graziani with his vastly superior armor could, if he chose, drive all the way to the Nile, and the British had somehow to delay him until their much-needed troops and tanks arrived. But how could it be done? Wavell turned for the answer to Colonel Dudley W. Clarke, an imaginative officer who had helped create Britain's famed commando teams and was now the director of A-Force, the Middle East branch of the LCS. Clarke, with Wavell's approval, decided that since the necessary reinforcements hadn't yet arrived they would simply have to be created on the spot.

Magic, military or otherwise, is best achieved by magicians, and at this point Major Jasper Maskelyne enters the camouflage picture. Traditionally, wizards make their appearance in a large puff of smoke, but in Maskelyne's case it was gunsmoke. Prior to the war he had been a well-known conjuror who traveled with his family's theatrical company, Maskelyne and Devant; and had subsequently received training at Britain's Camouflage Center at Farnham. With his elegant profile, bulldog chin, and trim pukka moustache, Maskelyne looked more like an elite product of Sandhurst than of England's music halls, but his professional expertise fitted him perfectly for his new military role. All through the conflict—at varying times in Britain and France, in Sicily, Malta, and Italy, in Syria, Cyprus, Egypt, and India—wherever a special combat trick or bit of visual chicanery was needed, the shadowy illusionist was usually involved.

Now, under Clarke's direction, a ruse was developed and put into operation by Maskelyne and his versatile camoufleurs, whom he referred to affectionately as his "Magic Gang." Using techniques that would often be repeated later—and always with effective results—they built dozens of dummy Cruiser tanks and heavy field guns with scrap materials rushed from Cairo to the Sidi Barrâni front. Attached to each "gun" was a length of drain pipe filled with chemicals that produced realistic loud flashes. Maskelyne's "gun flash recipe," as he described it, was: "Four teaspoons of black powder, six dessert spoons of aluminium powder, one teaspoon of iron filings. The first provided the smoke, the second the flash, the third the red flame."[6] The camoufleurs also simulated roads and tank tracks leading south from the coast toward the sector where Graziani's victorious divisions were camped, and the decoys were set up throughout this area. The next step was to add a believable air of turmoil and activity, and for this the Magic Gang hired a platoon of natives who drove their camels and horses across the dunes dragging wooden frame devices behind them; these stirred up satisfying clouds of desert dust which, seen from a distance or from high in the air, appeared to be maneuvering tank columns.

The Italians, equipped with reconnaissance planes, came over to look at and photograph these developments, but British antiaircraft fire kept the planes too high to uncover the precarious make-believe. Graziani knew from his intelligence bulletins that the British were expecting heavy reserves. Never a very aggressive general, he studied his photos and reports unhappily, concluded that Wavell was building up a huge force on the Italian right flank, and ordered his troops to dig in and go on the defensive.

The British, shifting their decoys, stirring up clouds of desert sand and periodically firing their noisy "barrages," were able to maintain this deception for many weeks until their reinforcements arrived and a real offensive buildup could begin secretly. Then on December 9, though still outnumbered, the British Eighth Army counterattacked; and within the next two months advanced 650 miles into Libya. In that drive Wavell's troops won an overwhelming victory, destroying 9 Italian divisions and taking 130,000 prisoners, 400 tanks, and 1300 guns, with minimal losses to themselves.

Maskelyne's work was finished for the moment, but the dominance of the Army of the Nile was to be short-lived. On February 15, 1941, Rommel arrived in North Africa to take charge of German troops in Libya. His orders were to provide "support" for Graziani's crumbling Italian armies, but the Wehrmacht commander had far more aggressive plans. A brilliant master of armored tactics, Rommel was also a camouflage enthusiast; and as soon as he reached Tripoli, one of his first orders was for the construction of dummy tanks in large numbers. When these decoy forces were deployed, the general had small airplane propeller engines attached to the backs of trucks, and as the trucks rolled across the desert the props stirred up a lively wake of sand that resembled the wakes of Mark III panzers. Rommel also ordered colored signal flares set at night to indicate spurious landing strips, troop concentrations, and supply areas. Another of his favorite tricks—which he was shortly to employ—was to use captured British trucks and carriers in battle. This not only helped to build up his limited transport, but sowed confusion among his opponents during the fighting.

The Desert Fox, without waiting for his own reinforcements—the 15th Panzer Division—to reach the front, struck back at the British toward the end of March. Wavell's forces had been badly thinned out in order to rush troops to help defend Greece, and now they were caught off balance. They also faced an unexpected new weapon: the German 88 mm. gun, which was designed originally for use against planes, had been adapted as a deadly and devastating antitank weapon. The British, outflanked and outfought, were soon reeling back into Egypt, and by mid-April all of the Libyan coast was in Rommel's hands except for the fortress town of Tobruk, which the panzers had bypassed.

With the Afrika Korps on the Egyptian border both sides paused to regroup and re-equip and Tobruk, ringed with Rommel's steel, was placed under siege.

At that time there were only five trained British camouflage officers in all of North Africa, their work coordinated by Major Geoffrey Barkas, a former film producer who had served in World War I and re-enlisted in 1939. Barkas had volunteered to serve in any useful capacity, little suspecting that his eventual mission—as camouflage director for the Western Desert Forces—would be to hoodwink the craftiest and shrewdest of all Hitler's generals.

North Africa soon became a testing ground for the tricks

and inventions that later paid off so handsomely in the Normandy invasion; and before long Barkas was joined by many other camouflage experts including Peter Proud, art director of British films; Derek Von Berg, a Johannesburg architect; Stephen Sykes, painter and artist in stained glass; John Codner, the painter, and E. C. Galligan, commercial artist; plus the redoubtable Jasper Maskelyne, who had gone off to Cairo with his Magic Gang to work on various tactical experiments and illusions. But the most urgent problem for the British at that point was the besieged coastal enclave of Tobruk. As Major Barkas noted in his memoirs:

> The enemy had almost complete air superiority and practically every fold and corner of the fortress was under constant observation and attack by guns and aircraft. . . . Certain places and equipment were vital to the garrison's capacity to hit back and even to survive, and by one means or another the enemy's fire had to be dispersed or diverted from these crucial targets.[7]

Tobruk could only be supplied by sea, and to keep this lifeline open, there were anxious demands for ruses of every kind. The town's modest harbor, with lighters and supply ships coming and going, was a wide-open target, and every anchored ship was exposed to Rommel's deadly dive bombers; but the camoufleurs followed the sound principle of using their terrain. In this case the terrain was the harbor, and in it were numerous wrecked and half-sunken ships; so the deception squads expanded this convenient wreckage with canvas and scrap metal, creating covered areas where small boats could be safely moored. The additions to the wrecks were so artfully designed that Nazi recon pilots failed to notice any major change in their appearance; and under cover of darkness, lighters were brought in and quickly hidden under the jumbled, rusting canopies. The success of this strategem was soon apparent since Luftwaffe bombers continued to attack unconcealed vessels, while the hidden ones (and their precious cargoes) escaped destruction.

For months during that siege, the only British aircraft inside Tobruk were three overworked Hurricanes at a small landing ground. Two of these were secreted by Barkas's camouflage crews in caves dug into a nearby wadi, with the entrances hidden by scrim nets and painted cloth. The third plane was literally buried in a large pit dug during the night alongside the landing area. The pit was covered with a giant wooden lid on which sand was spread during the day, and when the aircraft was to be used it was hoisted from the pit with a winch and cable. To reinforce this concealment, a decoy airstrip with several creditable dummy planes was built some distance from the real field. For over four months the Hurricane pilots were able to fly unimpeded, and Barkas reported that ". . . it was amusing to discover from annotated air photos in the possession of a captured enemy airman, some time later, that the dummies were marked as real, and that there was no indication of the real aircraft or their concealed hangars."[8]

Another problem at Tobruk was the distillery plant, which supplied a large proportion of the garrison's drinking water. The British knew that it was only a question of time before this was attacked, but hiding the installation or creating a decoy was pointless since the Italians, who had built it originally, knew its exact location and also understood its vital importance to the defenders. Finally Captain Proud, in charge of protecting the distillery, decided on a form of deceptive judo that would turn the enemy's aggression to advantage: A special squad of "wreckers" was quickly trained and hidden near the huge target and they didn't have too long to wait. On a clear moonlit night, a flight of Junkers came over the distillery and dropped scores of bombs. Fortunately their bombs landed close to the target, but not close enough to do real harm; and as soon as the smoke and dust settled Proud's "wreckers" went to work, following the same methods used by

Colonel Turner's Department in Great Britain. They dug numerous shallow "bomb craters" among the buildings, accenting them with shadows made of oil and coal dust. Debris was scattered widely. On the roof of the main works, the squad simulated other damage with canvas, paint, and cement. As a final touch unused parts of the station, including a large cooling tower, were blown up with preset charges.

By sun-up the magicians had completed their work and soon afterward—on schedule—Nazi photo planes, flying above British AA fire, came over to record the results of their bombing mission. The ensuing pictures showed bomb damage everywhere, the cooling tower in ruins, large black jagged holes torn in the roof and sides of the main building, and smoke still rising from the dismal wreckage.

The next military communiqué from Axis Headquarters in Rome proudly announced direct hits on the Tobruk distillery, and for weeks afterward the "wrecked" plant produced fresh water without enemy interference.

During all this, both sides had been urgently rebuilding their strength. Sir Claude Auchinleck had replaced Wavell as commander of the British Eighth Army and now planned a major offensive, code-named "Operation Crusader," to relieve Tobruk and drive the Afrika Korps back into Cyrenaica. The British buildup, involving the movement of vast tons of fuel, ammunition, and supplies to the forward areas, was carried out with much secrecy. As part of the cover plan, a dummy railroad line was extended *away* from the genuine railhead, ostensibly to serve as a special "tank-delivery spur." An Engineer company under Captain Stephen Sykes created this complex decoy, complete with dummy camps, cookhouses, shelter trenches, supply points, even a dummy train consisting of thirty-three freight cars, eighteen flat cars and one respectable-looking locomotive which—with the help of an old camp stove—belched great quantities of impressive smoke from its cardboard funnel. Miles of fake railroad track made of tin fuel cans (cut and hammered into proper form by Italian prisoners of war) were also put down across the windswept sands.

During the preparation of this ruse an unexpected crisis occurred. Wood for carpentry was virtually nonexistent in that area, and the decoy train had been built of extremely light materials, mostly woven rush mats which were then covered with painted cloth.

"A few weeks before the battle," Stephen Sykes recalled in a recent interview, "a bad sandstorm came up. The wind was really fierce, and the locomotive just took off. It just went up and sailed away, and kept on going for miles. We finally had to send out a search party to find it and fetch it back again." Fortunately the sandstorm also grounded enemy observation planes, and the locomotive was located, patched up, and rushed back to its tin rails before the hoax was unmasked.[9]

This decoy tank spur was successful in drawing fire away from the genuine buildup, and over one hundred enemy bombs were wasted on it, greatly easing the pressure on the real British railhead at Cappuzzo. Sykes still recalls the intense pleasure he felt as he watched the bombs whistle down on his absurd "train," justifying the arduous and painstaking work of the camoufleurs.

The Crusader offensive, launched in mid-November, gained the desired surprise; the battle was unremitting but the British, under Auchinleck's skilled generalship, finally lifted the siege of Tobruk and swept Rommel's panzers westward to El Agheila, on the Gulf of Sidra.

Curiously, during the confused fighting, while the German tanks were attempting a brief counterattack, their columns passed within a mile or two of the British army's main supply dumps which were jammed with fuel, food, and other vital stores. If the Germans had seized or destroyed these, Auchinleck's drive would have been seriously crippled; but the enemy's scout cars and recon planes *never even noticed them.* The superb camouflaging of these huge depots, each of which was several miles square, showed the deft and unmistakable hand of the talented Maskelyne, who had transformed thousands of tons of military supplies into innocent scrub patches and sand dunes. After the war, when General Fritz Bayerlein, Rommel's chief of staff, learned about the hidden supply dumps from Desmond Young, he was dumbfounded. Thinking of the valuable stores—particularly the tons of tank fuel lying under their very noses—the astonished Bayerlein told Young, "If we had known about those dumps, we could have won the battle."[10]

But Rommel was highly resilient, his setback here was temporary, and a shattering Axis victory was soon to come. New supplies were being rushed to the Afrika Korps by convoys crossing the Mediterranean, and the Germans and their Italian allies launched hundreds of raids against the great British base at Malta to protect these ships and drive off the RAF. In December 1941 and January 1942, they bombed Malta 430 separate times, and under cover of these heavy attacks the Axis supply convoys slipped through.

Rommel now proceeded to turn the tables and dupe the British. At a secret meeting of his top staff officers, he ordered them to spread discreet rumors that the Afrika Korps was preparing to pull back again into Tripolitania. The rumors spread quickly from El Agheila to Benghazi, to Tunis, to Bizerte, to Alexandria, to Rome itself: "Rommel is pulling out!" When the news reached General Auchinleck in Cairo he was, of course, highly dubious, and ordered extra visual and photo reconnaissance, but the recce missions tended to confirm the rumors. The Axis front was extremely quiet, with no evidence of military buildups or offensive preparations. Rommel was naturally making good use of camouflage. All movement and transport took place at night, supplies were well hidden under nets, and to insure secrecy, none of his unit officers received their attack plans until the very last moment.

To enhance his hoax, Rommel ordered demolition activities behind the German lines at Mersa el Brega. Many houses (all empty) were set on fire and many ships in the harbor (all useless hulks) were blown up. British secret agents dutifully notified Cairo that the Germans and Italians were destroying supplies and strongpoints before beginning their withdrawal.

While Auchinleck and his staff still remained skeptical, these reports (coming on the heels of the British victories) tended to lull the Eighth Army into a misplaced sense of security; then on January 21, the highly improbable took place: Rommel, with fresh tanks, new self-propelled heavy guns, and extra supplies of fuel, attacked in force. Using every technique of ambush and misdirection, the Nazi commander slashed through Auchinleck's lines with his panzer brigades. The Tobruk garrison, which before had held out so stubbornly, fell quickly to the Germans who captured thirty-three thousand prisoners. Soon the entire Eighth Army was in full retreat, falling back in chaos toward Alexandria; and by the end of June Rommel's forces were menacing the El Alamein line where the British had dug in for a final desperate stand. The Desert Fox, who had been made a field marshal by Hitler as a reward for taking Tobruk, was now only sixty-five miles from Alexandria; and beyond lay Cairo, the Suez Canal, Palestine, and all of the vulnerable Middle East, waiting to become the Wehrmacht's next great prize.

By this time, however, Rommel's supply lines were badly overextended, his panzer units decimated, his troops suffering from dysentery and jaundice and above all, he was again short of fuel. So the two armies drew to an exhausted halt, facing each other across the sands of Alamein, to await what would be the final and most decisive of their desert battles.

Meanwhile Maskelyne and his group, now based at Abbassia near Cairo, were busy with other deceptions. Four months earlier the Japanese had overwhelmed Singapore and in the process destroyed two large British battleships, the *Prince of Wales* and the *Repulse*. They followed this victory with a surprise task force raid in the Indian Ocean, sinking two British cruisers and the aircraft carrier *Hermes*. As a result the Royal Navy was seriously weakened in the Pacific area, and Admiral Sir Andrew Cunningham was instructed to rush units of his Mediterranean Fleet to the Far East. This in turn left Cunningham without enough strength to protect the Egyptian coast from the fast, modern ships of the Italian Navy. Somehow, magically, new vessels—preferably submarines—would have to appear quickly. A large fleet of submarines based in British North Africa would certainly inhibit the Italians—the only hitch was that very few submarines were available; so orders from Cunningham went to Major Maskelyne to pull them out of his conjuror's hat.

All the admiral modestly required, Maskelyne recalled, ". . . was a fleet of dummy submarines, full size, able to float like real submarines, but also able to be folded up by a few men so as to travel in a five-ton truck."[11] It was essential, Naval Headquarters pointed out, that the dummies could be dismantled, loaded, transported for miles to a new harbor, and "launched" again, all in a single night.

Materializing a large submarine was far more demanding than anything Houdini ever attempted, and even the unflappable Maskelyne was staggered by the problems involved, but he and his crew went gamely to work. Using empty oil drums, pipes, cable, canvas, and paint—plus the welcome remains of several wrecked railroad cars—they managed to simulate a full-scale British submarine 258 feet long and 27 feet high (not counting the periscope), plus the usual deck gun, anchors, and other paraphernalia. The remarkable vessel, floating on railway ties, was built at a secret beach on the Egyptian east coast, and was so realistic that RAF pilots began reporting the mystifying presence of an unidentified sub (possibly Japanese) in the Red Sea. Much alarm and confusion resulted, until the local air commander received a clarifying teletype from the British Naval Staff:

> REGRET MISUNDERSTANDING BUT MASKELYNE YOUR AREA—
> OTHER TRANSFORMATIONS WILL BE TAKING PLACE.[12]

Four such dummy subs were created by the Magic Gang, and were eventually moved to the British submarine base at Beirut. The displays helped to pad out the "sub count" being carefully kept by Axis recon planes, and also took the place of real submarines when those slipped out to sea on covert missions. Later, one of the decoys received the true camouflage accolade: It was attacked and sunk by a German dive bomber, whose pilot was undoubtedly hailed as a hero when he returned to his air base on Crete.

But not all of the major's hoaxes met with success. Cunningham's aides were so pleased with the ersatz subs that they gave Maskelyne an even more difficult assignment. On one of the Suez lakes was an ancient rusted British cruiser—already considered obsolete in 1918—which had been moored there to end its days quietly as an auxiliary flak ship. Why not, the Naval Staff wondered, revive the old warrior, give it a modern new "superstructure," and use it on naval maneuvers to mislead the Italians?

Maskelyne, with great misgivings, agreed to try the deception. "I went to look at it," he recalled, "and was told that I must never send more than three men aboard simultaneously, as otherwise the weight would cause it to turn turtle."[13]

The venerable hulk was shored up with underwater booms built of oil drums and wooden scaffolding; then the camoufleurs went to work to convert it into a menacing, heavily gunned British battleship 650 feet long. With canvas and chicken wire, paint, pipes, and perspiration, the Magic Gang actually achieved a version that could under certain circumstances—in very hazy weather—pass for the real thing. Unfortunately, Naval Ops was so enthusiastic that it was decided to tow the decoy out into the Mediterranean, although as a worried Maskelyne pointed out, ". . . the crazy additions we had built on were never intended for open water."[14]

The major's fears proved valid. During towing operations the convoy hit a patch of rough weather and the ancient vessel, buffeted by the waves, promptly sank with its flags flying. There is no evidence that Maskelyne felt the urge to adhere to tradition and go down with his doomed ship.

After these seagoing ventures, the magician went back to his experiments at Abbassia with collapsible tanks, trucks and guns which would not only look authentic, but could be packed up quickly and transported en masse to any part of the front. Fortunately there was a large pool of native labor available, which simplified the problems of quantity production. The camoufleurs also created platoons of dummy soldiers to man their fake guns; enemy pilots were growing wary, and one giveaway of a decoy artillery post was the absence of an adequate crew, so Maskelyne's straw men in battle-dress, plus a sprinkling of real soldiers, gave the sites plausibility.

Reflecting on his bizarre activities—the teaspoons of "gun flash" powder, the collapsible tanks and trucks, the sham submarines and all the rest—Maskelyne wrote in retrospect, "It was hard, at times, to remember that it all had such deadly meaning."[15] In this he echoed the feelings of many wartime conjurors, but the meaning of their work was, of course, terribly deadly, and their various sorceries brought practical and tangible results.

Experience gained in these seesaw battles on the rim of Africa also prepared the camoufleurs for one final hoax, which Geoffrey Barkas appraised as ". . . a visual deception on a scale unsurpassed and probably unequaled in British military history up to that time."[16] The mammoth deception had one crucial goal: to help put an end at last to Rommel, the Afrika Korps, and the ominous German threat to the Middle East.

And now, at El Alamein, this fantastic desert con-game was about to unfold.

12

Bait for the Fox

1942–1943

Summer of 1942—the summer of Rommel—saw the Allies face-to-face with calamity, and the fortunes of the British in North Africa, despite their best efforts, were at their lowest ebb.

During its advance the Afrika Korps had captured sixty thousand prisoners—British, South African, Australian, New Zealander, and French—as well as two thousand vehicles and mountains of provisions; Rommel, now at the peak of his professional career, stood at the gates of Alexandria and the Nile Delta, and one final offensive might conceivably carry him on through the Middle East to an eventual linkup with German armies driving into southern Russia and the Caucasus. It was a grim prospect which the Allies—Churchill in particular—were determined to prevent.

In mid-August after a flying visit to Cairo, the prime minister relieved Auchinleck, General Harold Alexander became C in C of the Middle East, and a comparative unknown named Montgomery was made commander of the British Eighth Army. Over five thousand trucks were hastily rushed to Egypt from America's assembly lines, as well as hundreds of powerful new U.S. Sherman and Grant tanks; and fresh troops poured in to take the place of those lost in the recent retreat.

The Malta garrison had by then been reinforced, and British ships and planes hammered Italian convoys carrying oil and munitions to the Afrika Korps. Mussolini, responsible for supplying Rommel, had grandly promised him great quantities of vital fuel, but the Allies—in a brilliant intelligence coup known as "Ultra"—had broken the Axis secret codes, knew all about the Italians' convoy plans, and sank their ships systematically.

"In September," Churchill later wrote, "30 percent of Axis shipping supplying North Africa was sunk, largely by air action. In October the figure rose to 40 percent. The loss of petrol was 66 percent. In the four autumn months over 200,000 tons of Axis shipping was destroyed."[1]

Meanwhile the Allied buildup continued, and Montgomery's forces soon outnumbered Rommel's two-to-one; in the air the ratio was three-to-one, and the British, with almost a thousand tanks, had more than twice the number available to the enemy. But Montgomery, his chief of staff, Brigadier Francis de Guingand, and his other officers knew better than to rely on numerical strength alone. Rommel was far too dangerous—and desperate—a tactician to be treated casually; and he could in one lightning move, as he had done in the past, overcome any military imbalances.

At that critical point the Egyptian front was a relatively narrow one of some thirty-five miles, its northern end anchored just west of El Alamein on the Mediterranean Sea and its southern terminus abutting a vast and impassable desert bog known as the Qattara Depression. On this cramped line there wasn't much room for daring sweeps or complex maneuvers, so the German options were limited; also, time heavily favored the British who were daily growing stronger. The flinty Rommel was, of course, aware of this and knew that he would have to make his move quickly or not at all.

But the British—with the help of Colonel Clarke of A-

Force—were ready this time for the depredations of the Desert Fox. For months Rommel had been receiving secret intelligence from a team of Axis agents operating near Cairo, known as the "Kondor" mission.[2] A British security unit under Major A.W. Sansom finally rounded up this group, and succeeded in learning their cipher system which was keyed to passages from Daphne du Maurier's novel, *Rebecca.* With this code in their hands, A-Force operators were able to send faked messages from "Kondor" via the spies' German contact in Athens. These, forwarded to Rommel, indicated that Montgomery was preparing to make his final stand along a ridge south of El Alamein called Alam Halfa, but (the reports added) he was still awaiting reinforcements and wasn't ready for more than a very makeshift defense. Rommel is said to have been overjoyed on receiving this "secret information" from his trusted source, and made his battle plans accordingly.

Clarke and de Guingand had still another bit of bait for the German trap. On one dark August night, reported historian Paul Carrel, ". . . the sound of fighting came from the German minefield on the southern front. Mines exploded and German sentries gave the alert. Machine guns were trained and flare pistols fired. . . . Apparently a British reconaissance troop was taking back its wounded."[3]

A German patrol was quickly sent out to investigate. They found a wrecked British scout car, bloodstains on the ground, and some hastily abandoned equipment inside the vehicle, which had obviously run over a mine. Among the scattered items was an officer's map case, also bloodstained, and in it an annotated map. At Wehrmacht Forward Headquarters, the captured map was carefully examined. The area around Alam Halfa ridge, with its drifting dunes and half-hidden wadis, was fairly treacherous. Some sections, where the sand was firm and solid, were known in military terms as "hard going." In other places the terrain was soft and impassable, and vehicles could quickly bog down in them. The captured map showed the "hard going" routes and danger spots in clear detail; much more detail than Rommel's cartographers had been able to provide. Well marked with serial and code numbers, creased and worn, frayed around the edges, stained with rings from numerous late-night tea cups, the map looked authentic. The Germans were cautious, but after checking and re-checking it against their own charts, the verdict of Rommel's staff officers was that their booty was genuine and of considerable value.

The map was, of course, a complete fraud. Clarke's artisans had prepared it to show "hard going" where there wasn't any and impassability where the routes were sound, and its unfortunate "loss"—explosion, wrecked car, bloodstains and all—had been staged by the British with dramatic effect.

The results weren't long in coming. Anxious to strike quickly, and basing his tactics partly on the captured map, Rommel launched his attack on the night of August 30. The British—again helped by intercepted code messages from "Ultra"—anticipated him and were waiting with strong forces dug in and well camouflaged; and German tank crews and infantry were met by a withering, unexpected fire, much heavier than they had thought possible. By dawn Rommel had failed to gain any of his planned objectives, and now his tanks and half-tracks (guided by the map) began to run into soft sand where they floundered in confusion. At this point, squadrons of Hurricanes and Bostons came over to strafe and bomb the helplessly bogged armor. For several days Rommel tried desperately to salvage his crumbling offensive, but the situation grew untenable and on September 4 he ordered a general withdrawal to safe positions behind his own minefields.

The brief and bloody battle, in which the Germans lost forty-eight hundred men, fifty tanks, and seventy heavy guns, had been a disaster for the Afrika Korps; Rommel's offensive strength was crushed; and now it was the turn of the revitalized British.

Earlier, soon after Montgomery had taken over direction of the Eighth Army, Geoffrey Barkas—now a lieutenant colonel—had been ordered by Clarke to report to Brigadier de Guingand's command caravan at Burg el Arab, behind El Alamein. He arrived with his aide, Captain Anthony Ayrton, and the two camoufleurs received top-secret instructions which left them both stunned. The long-planned British attack on Rommel's position, the chief of staff informed them, would take place along the northern sector of the Alamein line; but to mislead Rommel and gain surprise the British were going to falsify an enormous buildup at the southern end, near the Qattara Depression. This involved a double deception: The vast buildup in the north had to be hidden, disguised, minimized; and the fake buildup in the south had to be organized on such a large scale and with so much veracity that the normally crafty, intuitive field marshal and his staff would be completely taken in. Besides the positional hoax, there was also a time ruse: The fake buildup was to be arranged in such a way that Rommel would believe he still had extra days before the British attack would be launched.

The director of camouflage, overwhelmed at first by the scope of this ambitious plan, was plagued by worry and self-doubt, but he accepted the challenge. For Barkas and his camoufleurs, the new assignment was a truly staggering one, since Montgomery's immense force by then totaled eighty battalions of infantry, thousands of vehicles, and over six thousand tons of stores—all of it spread over a flat, highly visible area marked by a few stony ridges and some low camel-thorn shrubs. Cave Brown, who interviewed de Guingand after the war, notes that the brigadier confided privately to Dudley Clarke, "Well, there it is. You must conceal 150,000 men with a thousand guns and a thousand tanks on a plain as flat and as hard as a billiard table, and the Germans must not know anything about it, although they will be watching every movement, listening for every noise, charting every track. . . . You can't do it, of course, but you've bloody well got to!"[4]

Montgomery's offensive, "Operation Lightfoot," was scheduled to begin on October 23, and Barkas and his camoufleurs were given just *one month* to plan, prepare, and carry out their sleight of hand. The unprecedented scheme, to be coordinated by Clarke and supervised in the field by Barkas, was code-named "Plan Bertram," a highly prosaic title for the inspired visual poetry soon created by the illusionists among the sand dunes of the Alamein front.

It must be noted at this point that all the Allied hoaxes in North Africa (and elsewhere) were prepared with the greatest possible secrecy and no detail was overlooked; nor was their success due to German innocence and gullibility. Rommel was celebrated for his sixth sense, or *Fingerspitzengefuhl,* a cumbersome term for "intuition in the fingertips." He had a superb instinct for a combat trap or ambush, and could react with devastating rapidity. In view of this, how were the Germans so consistently fooled by the Allies? And how, in turn, were the Allies often misled by enemy deceptions?

The reasons lie in the dynamics of battlefield tactics. Both sides knew that decoys and camouflage illusions were being regularly used, but even with this knowledge, doubts and uncertainties remained; and cross and double-cross became so much a part of the North African deception game that no commander could afford to make dangerous wrong guesses. During the relief of Tobruk, Maskelyne wrote, "The enemy spies and reconnaissance experts knew we had dummy tanks and guns—but could never discover which was which, the dummy or the real." Given the fluid complexities of desert warfare—where "real" tanks might turn out to be wooden fakes and supposed "decoys" suddenly begin spitting high explosive shells—neither side could relax its guard. So Rommel, now on the defensive, had to maintain vigil wherever there was *the slightest doubt,* and his resulting uncertainties led to patterns of thinking which Operation Bertram would try if possible to exploit.

Colonel Barkas was assisted in this scheme not only by Tony Ayrton but by other camouflage experts such as Majors

R. J. Southron and V. W. Hampton, Captain Phillip Cornish, and Lieutenants Brian Robb and Sidney Robinson. To help mass-produce the necessary battalions of dummies, three Pioneer companies were assigned to Bertram. Later, native labor in Cairo and Alexandria was also used; but to guard the project's secrecy, the civilian workers were given unidentified components to produce without knowing what their purpose was or how they would be used by the unfathomable British.

The hoax divided itself essentially into two parts: *concealment* in the north and *display* in the south. The first challenge facing the conjurors was to secretly shift some six thousand tons of supplies to the El Alamein area, within five miles of the northern attack line. Crucial among these supplies were several thousand tons of tank fuel; and here the camoufleurs made a lucky discovery. Colonel Barkas recalled:

> Ayrton and Robb found that quite close to the Alamein [railroad] station there were one hundred sections of slit-trench beautifully lined with masonry. . . . They had been there for about a year, and by now the enemy must have come to regard them as part of the furniture of the battlefield. By experiment it was found out that if each wall of the trenches was given an extra facing of "masonry," in the form of petrol cans, the shape of the internal shadows remained virtually unchanged as seen from the air.[5]

At night, two thousand tons of fuel cans were neatly layered along the inner walls of these accommodating trenches, where they would be ready for instant use. As a test, RAF air observers were then invited to locate this secret cache, but were unable to find it. Fuel for Montgomery's armor was now safely in place.

The next problem was to transport and hide the vast quantities of food, ammunition, and other stores, and here a new visual stunt was tried. The crates and boxes were moved during the night into the assigned dump sites. Then they were stacked and piled to resemble the shapes of ordinary three-ton trucks. These truck-stacks were scattered about the area in much the same way as real trucks would be dispersed, and finally a large camouflage net was draped somewhat tightly over each stack, making the forms apparent. The illusion was all the conjurors had hoped for: Seen from the air, or viewed through binoculars from a distant ridge, the supplies looked like a fleet of ordinary camouflaged transport vehicles of no major combat significance. Overflow supplies were tucked under small tents nearby, resembling shelters for the vehicles' "drivers and mechanics."

Over four hundred 25-pounder field guns were hidden in forward areas in a similar manner. The camoufleurs found that by dovetailing each gun and its accompanying ammunition trailer, the whole unit would fit under a simulated "truck canopy" made of canvas hung from poles. Realism was heightened by the fact that the wheels of the gun carriages and trailers could serve as wheels for the spurious "trucks." Since each canvas frame neatly swallowed a complete artillery unit, the device was known as "Cannibal"—and soon hundreds of Cannibals had been quietly positioned north and south of Miteiriya Ridge.

The next—and greatest—challenge was to move up the many tanks and armored cars of Montgomery's 1st and 10th Armored Divisions, and these would have to be concentrated twice—first in the rear staging area, and just before the battle in forward positions facing Rommel's minefields. Here again hollow, collapsible frames, resembling British ten-ton lorries, were placed along the axis of the planned assault. The large wood and canvas shells, known as "Sunshields," were each designed to shelter a Sherman, Grant, Crusader III, Valentine tank or a half-track; and well ahead of time more than *seven hundred* of these had been manufactured, assembled, and moved into position under cover of darkness. The genuine armor, assembled some sixty miles behind the lines, was left relatively exposed as part of the plan. The British, with strong air superiority, were able to protect this concentration, but they wanted Rommel's recon pilots to observe that the heavy armor was situated well away from the front, thus posing no immediate threat. Several nights before the battle, each of these tanks would move up and go into hiding under a Sunshield. The tanks, dragging chainlike devices, would obliterate their own tread patterns; in addition, each tank as it moved forward would be replaced in the rear area with *an accurate dummy replica.* This meant that Rommel's pilots, continuing their daily check, could see Montgomery's "armor" still resting far to the rear, and—unwittingly counting Barkas's numerous substitutes—would assume that no sudden action was being planned.

Coordinating this giant maneuver called for superb planning and timing. The hollow Sunshields near the front were all placed in the precise positions later to be occupied by the real tanks and other vital gear. Each of these "hides" had a serial number, and was earmarked for a particular tank or particular unit of equipment. Every British tank crew was also given a corresponding serial number, taken up to the lines to see where its own Sunshield was located, and given instructions in how the device worked.

"It was a long and detailed job," Barkas wrote, "but when the order came to move up, each tank crew knew where to go, what to do, and how to get out of sight before dawn."[6]

It should be noted that the forward assembly point for Operation Lightfoot, called "Martello," covered a sector eight miles wide and twelve miles deep; and it wasn't unusual to have many thin-skinned or noncombat vehicles spread over an area of that size. Moving the Sunshield "lorries" into place ahead of time had certain other advantages. It was entirely possible that the Germans might discover these to be dummies, which would tend to strengthen their belief that activity in the north was merely a feint, and that the real blow would indeed fall at the southern end.

While all these concealment efforts were going on, other units under Cornish and Robinson were busy creating a false display of activity in the south. The keystone of this hoax rested on the ever-present need in the sunbaked Egyptian desert for fresh drinking water. An existing water pipeline was already in place, running south from the El Imayid railroad station, some miles behind El Alamein. The camoufleurs decided to misdirect Rommel in classic Houdini style by continuing this pipeline southward in dummy form.

Working with a crew of Engineers, the magicians began "extending" the pipeline five miles at a time, pacing themselves as they would for a genuine installation. Five miles of dummy "water pipe" made of the ubiquitous four-gallon fuel tins, were laid along a stretch of newly dug trench. At night the trench was filled in, the same five miles of fake pipe was advanced, and a new section of trench was dug. This process was repeated day after day, giving the impression of a line growing steadily southward to provide water for the mass of British troops who would later be assembling between the Munassib Depression and Mount Himeimat. At three points along this simulated pipeline—known as "Diamond"—dummy "pump-houses" were built; also overhead "tanks," "can-filling stations," and at one spot a large, shallow "reservoir." Dummy vehicles and soldiers were added, and regular truck traffic was rerouted by the Army so as to follow the pipeline and create further activity.

The Diamond ruse provided an extra subtle refinement, and perhaps the most important of all. The Germans had, of course, followed the gradual mile-by-mile progress of this installation, and could estimate the date at which it would reach its apparent terminus at Mount Himeimat. But the camoufleurs timed their work so that, by Montgomery's Zero Hour, they would still be several days *short* of reaching this objective, lulling the Germans again into believing that no attack was imminent.

To bolster their southern "buildup," the conjurors also created numerous dummy supply dumps and gun emplacements behind the Munassib front, manned by a scattering of real troops. The dummy depots, built under the direction of Brian Robb, simulated nine thousand tons of nonexistent supplies, spread out in hundreds of separate stacks. Robb and his native crews even built three small "administrative camps" to manage this giant "storage area."

Rommel, of course, watched all these developments with uneasiness. Normally, because of terrain conformations and the proximity of the shore road and rail line, he would have expected Montgomery to attack near Kidney and Miteiriya Ridges in the north; militarily it was the sensible option. But the heavy activity in the southern sector puzzled and discomfitted him. Were the British perhaps planning a strong feint in the north, and their real attack in the south? Or was it vice versa? Why were they expending so much time and effort to push a pipeline south, carrying gallons of precious, badly needed water, if not to support a major operation? Which were the British dummy forces and which the real? Where would the main blow land?

Fingerspitzengefuhl notwithstanding, the crafty desert fighter did what was in essence the logical thing, and made what was in fact a grave tactical blunder: He split his armor to cover *both* eventualities, keeping the Fifteenth Panzer Division and the Italian Littorio Division in the north and moving the Twenty-first Panzer and Ariete Divisions to cover the southern sector opposite Munassib.

Curiously, at the very moment when he should have been close at hand, Rommel left the battlefield and returned to Germany. He had been suffering for some time from jaundice and a painfully inflamed throat, and badly needed hospitalization; he also wanted to confront Hitler personally and plead for additional fuel and supplies for *Panzergruppe Afrika*. The slow progress of the British pipeline, the lack of action in Montgomery's rear staging areas, and reassuring wireless messages from his agents in Cairo (faked by A-Force) had convinced him that there was no immediate danger. So leaving the elderly General Georg Stumme in command, he flew off to meetings with the führer and then to a hospital at Semmering.

Shortly after 9:30 on the night of October 23, the sky over the El Alamein desert was suddenly set ablaze by a tremendous artillery barrage. Montgomery's gun crews, throwing off the Cannibal "hides" on their guns, had gone into action, supported by heavier guns massed to the rear. As soon as the shattering barrage lifted British sapper teams, well trained and briefed, moved forward to clear prearranged channels through the German minefields. The armor crews, who had waited all day under cover, knocked the Sunshields aside and hundreds of tanks rumbled forward. In the south, camouflaged field guns of General M. P. Koenig's Free French Division also opened up. The great northern offensive was on, catching the Wehrmacht generals completely by surprise.

To add to their confusion, reports quickly came in that a British amphibious force was attacking the Mediterranean coast *behind* the German lines. At Sidi Abd el Rahman, German troops could hear, out beyond a heavy offshore smoke screen, the sound of naval vessels, ships' engines, the clanking of anchor chains and the shouted commands of British officers as troops clambered into their assault boats. Stumme, alarmed, detached an important reserve regiment (the 90th Light) and rushed it to the threatened sector; he also ordered Luftwaffe bomber and fighter cover, though his meager air units were badly needed elsewhere.

The coastal foray—successfully drawing enemy defenses from the northern front—was an extra touch again showing the wizardry of Jasper Maskelyne. The British "amphibious force" consisted in fact of a handful of rafts towed close inshore by motor-torpedo boats. Flares and gun flashes were fired from the rafts automatically; and the noises of battle were spread by recordings played over loud amplifiers. An element of "olfactory" camouflage had also been introduced: Smoldering canisters on the rafts duplicated the acrid smell of

British naval engines. In all it was a minor diversion, but a highly successful one.

Toward morning of that opening battle, a flustered General Stumme set out in an armored half-track to survey the chaotic front, but his vehicle blundered into a British antitank ambush and the German commander succumbed to an apparent heart attack.

Rommel learned quickly of these developments and, though still far from well, rushed back to the critical front. "When he arrived," Desmond Young notes, "the battle was already lost."[7] The fighting would still go on relentlessly and savagely; but the Afrika Korps, outnumbered, outgunned—and brilliantly outwitted—was inevitably doomed. With his remaining ten thousand Germans and some sixty tanks—plus twenty-five thousand Italians who had little desire to keep on fighting—the Desert Fox conducted a stubborn, dogged withdrawal as he was pushed back relentlessly into Tripolitania. By then Allied armies under Eisenhower had made fresh landings in Morocco and Algeria, and the famed Afrika Korps was trapped between two giant pincer arms. Later *Panzergruppe Afrika* was hastily reinforced and on occasion—notably at the Kasserine Pass—showed flashes of its earlier power; but after the Alamein debacle, Rommel and the Wehrmacht were essentially finished.

Reporting later on El Alamein to the House of Commons, the British prime minister stated, "By a marvelous system of camouflage, complete tactical surprise was achieved in the desert. . . . the enemy suspected that the attack was impending, but did not know how, when, or where, and above all he had no idea of the scale upon which he was to be assaulted."[8]

General Ritter von Thoma, who had taken over command after Stumme's death, and who was captured during the fighting, was interrogated afterward by British intelligence officers; he acknowledged that Rommel and the other German generals definitely believed that the main attack would come in the south, opposite Munassib. Furthermore, the assault (launched on October 23) wasn't expected by the Germans until some time in November.

But the final summation of this vast and most remarkable of desert deceptions belongs appropriately to Colonel Barkas:

> Though none of us was so foolish as to think that it had been won by conjuring tricks with stick, string and canvas, we could at least feel that we had earned our keep. . . . It was good to feel that camouflage had helped to put the fighting men into battle on more favorable terms, and so to purchase victory at a lower price in blood.[9]

Paradoxically, the success of Bertram wasn't an unmixed blessing; every deception, no matter how well executed, revealed a little more of the Allies' game plan, sharpened German insights, and put added pressure on the illusionists. So each new scheme had to somehow be a bit more subtle and skilled than the last, and more difficult for the enemy to unravel.

At this point, with the ending of the Axis threat to the Middle East, the stage was set for just that type of subtle connivery: a cover plan which involved Sicily, a set of counterfeit letters, and a body washed ashore in neutral Spain.

13

Mincemeat Swallowed Whole

1943–1944

The White House—May 1943. Churchill and his staff were again in Washington, D.C., for a third series of high-level conferences on strategy and planning. One urgent matter to be discussed with Roosevelt and the American generals was the coming Allied invasion of Sicily from newly won bases in North Africa.

While the leaders reviewed their complex plans and problems, a stream of classified messages poured in for the prime minister from his own war headquarters deep below Great George Street in London. Late one afternoon, Sir Hastings Ismay stepped in and handed Churchill a communiqué which had just arrived from the decoding room. The top secret dispatch from England stated in its entirety: MINCEMEAT SWALLOWED WHOLE.[1]

The prime minister glanced at the slip of paper, smiled faintly, and drew on his *Romeo y Julieta* corona with renewed satisfaction.

Hidden behind the three cryptic words of that message (and it remained hidden for ten years) was one of the most novel subterfuges of the war. Though it had neither the grand strategic design of Fortitude nor the intricate complexity of Bertram, the plan produced spectacular results and constituted, in the opinion of military analysts, a strategem equal to anything ever attempted in history.

The iconography of camouflage takes many forms, not all of them involving paint, nets, lumber, and canvas. During World War II, under the rubric of "disguise and divert," came techniques of visual and aural deception, faked maps, false code messages and radio signals, impersonations, and endless varieties of decoys designed to mislead and confuse enemy planners. The hoax known as "Operation Mincemeat" also concerned a decoy, and it took the form of a human corpse.

Because of its bizarre character, the story of "Mincemeat" properly begins in the middle, at the grave of a British officer who apparently drowned when a plane carrying him to Algiers crashed at sea. In a cemetery on the outskirts of Huelva, Spain, there is a horizontal slab of plain white marble, and on it is the inscription:

WILLIAM MARTIN
Born 29th March, 1907. Died 24th April, 1943.
Beloved son of John Glyndwyr Martin
and the late Antonia Martin of Cardiff, Wales.
Dulce et decorum est pro patria mori.
R.I.P.[2]

After his interment the grave of William Martin, a major in the Royal Marines, was visited frequently by various officials including the naval attaché in Madrid and the British vice-consul at Huelva. Floral wreaths were placed on the grave—loving tributes sent by grieving friends and relatives in England. In all, it was a fitting farewell to a man who had died in the line of duty far from home; but oddly enough none of the graveside visitors had ever seen or heard of Major Martin, and his "bereft" friends and relatives at home had never heard of him either. The soldier beneath the white marble slab was, in actuality, a complete fabrication; in real life he simply never existed.

William Martin, "the man who never was," had been conjured up by a team of British officers as part of a macabre masquerade carefully designed to dupe enemy intelligence and the German High Command. The originator of the scheme and its principal architect was Lieutenant Commander Ewen Montagu, then serving at the Admiralty in the Department of Naval Intelligence. Montagu's human decoy was created virtually out of thin air, but the need for him was real and tangible. The Allied invasion of Northwest Africa, "Operation Torch," had been launched by Eisenhower in November 1942. Coming a short two weeks after Rommel's defeat at El Alamein, the surprise landings at Casablanca, Oran, and Algiers were successful; in the months of hard fighting that followed the Allies took Tobruk, Benghazi, Tripoli, Tunis, and Bizerte, and the spring of 1943 saw the final collapse and surrender of all German and Italian forces in Tunisia. On May 13 of that year, General Alexander cabled Churchill:

> Sir: It is my duty to report that the Tunisian campaign is over. All enemy resistance has ceased. We are masters of the North African shores.[3]

Now powerful American and British armies were poised along the southern rim of the Mediterranean from Tangier to Port Said, ready to drive into the proverbial "soft underbelly" of Europe, but where would this next blow fall? At Rastenburg Headquarters, Hitler studied his maps and brooded as always over his territorial imperatives. The Allies, he felt, might strike anywhere—at Greece, Sardinia, Corsica, Sicily, Italy or southern France. He worried particularly about the Balkans; that area with its vast stores of copper, bauxite, chrome, and oil would surely be a tempting military objective.

Actually the Allies had already decided at the Casablanca conference in January 1943 that Sicily would be their next goal, and the campaign to take the island had been code-named "Operation Husky." Sicily, dominating the sea lanes in the central Mediterranean and lying just off the vulnerable toe of Italy, was a natural and obvious target—*too* natural and obvious, some of the strategists felt. Later, Churchill himself drily commented, "Anybody but a damned fool would *know* it was Sicily."[4] In view of that fact, Operation Husky would have to be shielded by a particularly ingenious cover plan. The necessary preparations for a major assault couldn't possibly be hidden from the Germans, but—as in other cases—they *could* be fooled as to its purpose and destination. Hitler was already worried about the Balkans. Could he be convinced, somehow, that the Allies were now planning *two* operations: one in the Greek islands as a step toward the Balkans, and the other against Sardinia as a jumping-off point for southern France? It was an interesting idea, but how could it be accomplished?

In a series of secret meetings at the Admiralty various plans were proposed and rejected; nothing seemed sufficiently workable. Then Ewen Montagu came up with a viable notion, or as he put it, ". . . the penny suddenly dropped."[5] Why not, he wondered, obtain a body, disguise it as a staff officer and plant high-level papers on the corpse describing fraudulent attack plans? To make sure the body fell into enemy hands, it could be washed ashore as though from a crashed plane. "He would float ashore with the papers," Montagu speculated, "either in France or Spain; it wouldn't matter which. Probably Spain would be best, as the Germans wouldn't have as much chance to examine the body there as if they got it into their own hands, while it's certain that they will get the documents, or at least copies."[6]

The plan was considered, explored, debated and finally (though with many misgivings) officially approved by the LCS. One major concern was that a deception of this kind had to be brought off without a hitch and every detail had to be impeccably arranged. If the Germans became the least bit suspicious the whole plan might backfire, and after they uncovered the hoax they could deduce the Allies' real intentions. Thus if the Nazis found that the "attack on Sardinia" was a calculated fraud, their attention would focus quickly on Sicily as a likely geographic alternative and the true target. Once alerted, they would have no trouble in finding *valid* evidence to support their deductions.

Montagu and his committee were painfully aware that Mincemeat was treading on dangerously thin ice and that every aspect would be examined meticulously by FHW and the *Abwehr* in an effort to find flaws. Conscious of the pitfalls, they went ahead with the first macabre task which was to locate an appropriate body, and for this they enlisted the advice of Sir Bernard Spilsbury, one of England's leading pathologists. With Spilsbury's help a covert search was begun and a suitable corpse was eventually found in a Westminster mortuary—that of a young man in his early thirties who had died of pneumonia and exposure. Since there would tend to be some liquid in the lungs of a pneumonia victim, this would fit smoothly with the notion of a man who had died from shock or drowning while floating in a rough sea. Permission to use the body was granted on one condition: that the committee never reveal the young man's true identity. Though there have been many rumors since then, Montagu's part of the bargain has been scrupulously kept to this day.

With the corpse placed in cold storage at a secret location, the next step was to give it a convincing identity, both military and personal. Experts at MI-6 fabricated the necessary credentials for "Acting Major William Martin, 09560, Royal Marines, a staff officer at Combined Operations Headquarters." These included a Naval Identity Card #148228 (with a photograph provided by a convenient look-alike), an official C.O. Headquarters pass issued and countersigned at Whitehall, and two metal identity tags stamped "Major W. Martin, R.M., R/C." The committee also invented personal items to give the major lifelike dimensions. Among these were a note from his "father," a crusty Victorian gentleman, and two love letters from his "fiancée," a charming young woman named "Pam." A photograph of Pam was supplied by one of the secretaries at the Admiralty, who agreed to its use after being told simply that it was needed in a top-secret deception.

Additional effects—which would be found in due time on Major Martin's body—included a bill for a diamond engagement ring from a Bond Street jeweler dated April 19, a pack of cigarettes (partly used), a box of matches, two bus tickets, assorted paper money and coins, a pair of ticket stubs for the April 22 show at the Prince of Wales Theatre, and a receipted bill dated April 24 from the Naval and Military Club in Piccadilly. The major, apparently careless with his money, also carried a stern reminder from Lloyds Bank pointing out that his account was overdrawn in the sum of £79.19s. 2., and informing him that if payment was not received forthwith ". . . we shall have no alternative but to take the necessary steps to protect our interests."[7] The dates on all these items were carefully planned so that when Martin's body washed ashore on or about April 30, it would appear that he had crashed on a flight from London, and had floated in the water for four or five days.

Next came the important documents that would be carried in a leather briefcase chained to Martin's wrist, and which could hopefully set the German bloodhounds on their false trail. Here Montagu and his conjurors had to deal with psychological as well as military factors; Germany's intelligence agents, shrewd and competent, would see through any letters or reports that appeared too obvious. The false information had to be fairly circumspect, with a few well-planted clues the Germans would have to decipher for themselves. This would strengthen the validity of the ruse by convincing the enemy experts that their findings resulted *from their own ingenuity.*

One of the key enclosures was a personal letter from General Archibald Nye of the Imperial General Staff to

General Alexander, Eisenhower's second in command at 18th Army Group Headquarters. His letter referred indirectly to a projected attack in the eastern Mediterranean on the Greek islands of Cape Araxos and Kalamata. He also intimated that an assault at the *other* end of the Mediterranean would be covered by a feint toward western Sicily. This diversion, Nye wrote, would be called Operation Husky. Here Montagu's committee added a unique refinement to the art of military misdirection. Husky was, of course, the name of the real operation against Sicily; but by implying that it was a deception, the Allies succeeded in covering their cover. Now, if other references to Husky accidentally turned up in Nazi intelligence reports, Canaris and von Roenne would assume that these dealt with an *unimportant feint* rather than the main attack, which would (so they thought) take place elsewhere.

A second letter placed in the precious briefcase was from Lord Louis Mountbatten to the commander of the Mediterranean Fleet, Admiral Andrew Cunningham, at Algiers. Major Martin was introduced as a young officer with special expertise in the use of landing barges and related equipment. In his note, Mountbatten wrote that Martin could be of great help in any forthcoming seaborne landings. He asked for the officer's return to Whitehall as soon as possible, and added, "He might bring some sardines with him. . . ."[8] The little joke, Ewen Montagu felt, was "frightfully labored," but the kind which would appeal to the Germans who would most likely grasp and understand the reference. Montagu was right; as it turned out, his heavy-handed jest about Sardinia played a helpful part in the later development of the scenario.

By now, Huelva had been pinpointed as Major Martin's final destination. The small city in southern Spain, on the Bay of Cadiz, had certain advantages. It was far enough from Gibraltar to discourage the Spaniards from shipping the corpse straight off to British territory for burial. Because of the prevailing tides and currents, it was also likely that a body left at sea would float accommodatingly into the Huelva harbor.

The committee knew too that a very efficient German agent was stationed at Huelva, and that he had excellent contacts with the local Spanish officials. If a British corpse drifted ashore, the German would soon have access to any papers found on it.

In due time Major Martin, properly uniformed, was carried aboard a British submarine, *H.M.S. Seraph,* laying off Greenock, Scotland. The corpse was inside a large metal cylinder packed with dry ice and labeled: "Handle With Care—Optical Instruments." Martin's personal letters and other belongings had been placed in various pockets, and the crucial letters, plus several other official memos, were inside the locked case which had been carefully chained to his wrist. Security was so tight that nobody aboard the submarine except for its commander, Lieutenant N. A. Jewell, knew the true nature of their secret operation.

The major's voyage to the coast of Spain proved smooth and uneventful. At 4:15 A.M. on April 30, the *H.M.S. Seraph* surfaced quietly off the Huelva shore and the canister was brought up on deck. The moon had set and the night was black. Directed by Jewell, a squad of four officers—all quite startled on noting the grim contents of the mysterious container—removed the body, fitted it with an inflated life jacket, and set it carefully into the water. One of the party wrote afterward, "A gentle push and the unknown warrior was drifting inshore with the tide on his last, momentous journey. 'Major Martin' had gone to war."[9]

Later the empty canister was taken farther out to sea, set adrift, and sunk with a short burst of machine-gun fire.

With the implementing of Mincemeat the work of British Intelligence was substantially complete, and the rest would be up to the Spanish and Germans; but the committee, standing by at Admiralty, didn't have long to wait. Shortly after sunrise on that same day, a Spanish fisherman spied the floating body and it was brought by launch into Huelva. A Spanish doctor examined the corpse at the local morgue and certified that the

British officer ". . . had fallen into the sea while still alive and had no bruises, and that death was due to asphyxiation through immersion in the sea . . . five to eight days before."[10]

On the following day Martin's body was duly handed over to the British vice-consul and was later given a funeral with suitable military honors. And now the British naval attaché at Madrid, Captain J. H. Hillgarth, began to play his assigned part in Montagu's drama. Obviously, if the whole episode were to appear legitimate, the British would be in an uproar about the leather case and its incriminating contents. Hillgarth, who had been secretly briefed on the deception plan, reacted with appropriate agitation. He wired the vice-consul at Huelva inquiring as to the whereabouts of Martin's official black briefcase with the royal cipher on it. Hillgarth stated further that the case contained several letters "of the utmost importance" and that he was "on no account to arouse the interest of the Spaniards in the documents." In reply to the vice-consul's discreetly urgent inquiries, the Spanish minister of marine stated that the briefcase and its documents would have to pass through "official naval channels," and that this would take several days. Pressure to obtain the papers was steadily increased by the British, politely thwarted by the Spaniards, and carefully (to the Admiralty's delight) reported to the Germans.

At long last, on May 13, the Spanish chief of naval staff turned the briefcase over to Hillgarth in Madrid with assurances that the contents were intact and "everything was there." Eventually the documents reached London where they quickly underwent scientific tests; the LCS had taken steps that would show if the letters had been read or tampered with and the results were gratifying. "There was little doubt," noted Montagu, "that the Spaniards had extracted the letters and knew what was in them, and . . . we could rely on the efficiency of the Germans to get all that they wanted out of that situation. We were sure that our confidence in the Spanish end of the German intelligence service would not be misplaced. It was now up to Berlin to play *its* part."[11]

Everything pointed to the fact that the deception had been a genuine success. Wishing to keep Lieutenant Jewell informed, Montagu sent him an ordinary picture postcard with the simple message, "You will be pleased to learn that the major is now very comfortable."[12] And it was at this point that the British Chiefs of Staff signaled "Mincemeat Swallowed Whole" to an anxious Churchill in Washington.

Exactly how well did Major Martin carry out his mission? The invasion of Sicily was launched on the night of July 9, and Eisenhower reported that, "Up to that moment no amphibious attack in history had approached this one in size." The assault, to be carried out by the U.S. Seventh Army under Patton and the British Eighth Army under Montgomery, was aimed at the southern coast of the island on a line running from Licata around to Syracuse. News correspondent Ernie Pyle sailed with Rear Admiral Alan Kirk's assault convoy, and as the great fleet headed across the Mediterranean toward the Sicilian beaches Pyle reported:

> We stood at the rails and wondered how much the Germans knew of us. Surely this immense force could not be concealed; reconnaissance planes couldn't possibly miss us. Axis agents on the shore had simply to look through binoculars to see the start of the greatest armada ever assembled up to that moment in the whole history of the world.[13]

At that time neither Pyle nor anyone else in the huge convoy knew the truth: The armada was certainly known to the Italians and Germans, but they had been led to expect a diversionary attack on the *western* end of Sicily—far from the southern beaches—followed by a major thrust against Sardinia to the north. The landings on the south coast took the Axis garrison by surprise, and Allied divisions poured ashore with a minimum of opposition. In his account Eisenhower wrote, "As battle reports began to arrive it was evident that the enemy had been badly deceived as to the point of attack. His best formations were located largely on the western end of the island, which he had apparently believed we would select."[14]

Commander Montagu's fabrication produced even more far-reaching results. From German military archives captured later at Tambach, it was clear that von Roenne and the other intelligence chiefs had indeed swallowed Mincemeat whole. The records show that they accepted the evidence, analyzed everything including Martin's personal effects with Teutonic thoroughness, and reported in detail to Hitler and OKW. Many top-level intelligence estimates were circulated. One carried a German translation of General Nye's letter along with the comment, "The genuineness of the captured documents is above suspicion,"[15] and it went on to say that there was little likelihood of Allied deception in the incident. Another report—vindicating Montagu's labored pun—noted among other things that "A joking reference in the letter points to Sardinia."[16]

Hitler apparently accepted these analyses, which by and large fitted his own beliefs; in an urgent directive to his commanders on May 12, he outlined the Mediterranean defense situation and concluded with the order: *Measures regarding Sardinia and the Peloponnese take precedence over everything else.*

The German High Command responded with its customary zeal. A crack SS brigade was added to the defense of Sardinia. A panzer division, stationed in France, was loaded quickly aboard 160 railroad cars and rushed to Greece. OKW detached two other panzer divisions from the crucial Russian front and sent them on a ten-day journey to bolster the Greek defenses. Naval forces were reassigned and one squadron of Kriegsmarine R-boats was shifted from Sicilian waters to the Aegean. During the following weeks, Greek minefields were enlarged and shore batteries were strengthened, while in the western Mediterranean Sardinia and Corsica were reinforced *at the* expense of the Wehrmacht's Sicilian defenses.

Mincemeat paid continuing dividends. On August 17, Patton's U.S. Third Division drove into the town of Messina on Sicily's northeast tip, just a few miles across the straits from the Italian mainland. "The man who never was" had surpassed everyone's expectations; Operation Husky succeeded as planned, and now all of Sicily was under Allied control.

William Martin, R.M., 09560, wasn't the only human decoy employed by the Allies during the course of the war. Impersonations, disguises, sham appearances, and other forms of corporeal camouflage were all used at times to support specific deception schemes. One of these plans involved Lieutenant Meyrick James of the British Army Pay Corps, who attained brief notoriety afterwards as "Monty's double." Early in 1944, an officer at SHAEF's Committee of Special Means saw a photograph of Lieutenant James and was struck with his remarkable resemblance to the noted British field marshal and hero of Alamein. The physical similarity was quite uncanny and the officer, Lieutenant Colonel J. V. B. Jervis-Reid, thought it could be of use in a novel deception.

Under direction of the CSM and Jervis-Reid, "Operation Copperhead" was soon activated. Copperhead's purpose was twofold. The Allies were then preparing for the D-day assault on Normandy, and the Germans were aware that Marshal Montgomery would play a key role in that operation. But if he were to suddenly appear in North Africa, the enemy might be led to believe that he was preparing to lead an attack from Italy and Africa against southern France. This could help to pin down four strong Wehrmacht divisions south of the Loire, away from the Normandy front. In addition, it might persuade the Germans that the cross-Channel invasion (for which Montgomery was needed) had obviously been postponed.

The ruse had definite possibilities, and James—who had been an actor in peacetime—seemed the ideal man for the impersonation. For weeks, the very surprised paymaster was drilled and trained in the British commander's gestures and mannerisms, his personality, and his way of speaking. Then, properly clothed in an exact duplicate of Montgomery's well-tailored battle dress complete with epaulets, decorations, and the inevitable black beret with its two badges, James was ushered aboard a Liberator bomber and flown to Gibraltar. After an overnight stay at which "Montgomery" was officially wined and dined (and his presence presumably noted by German agents) he was flown on to Algiers. James's appearance—heightened by a judicious touch of makeup and his excellent duplication of the commander's walk and bearing—indeed created an astounding effect; and British workmen, both at Gibraltar and Algiers, greeted his command car with shouts of "Good Old Monty!" and "Hurrah for Monty!"

In North Africa there were further official meetings and ceremonials, and "Monty" was heard several times referring to a mysterious "Plan 303." But at this point a serious hitch developed.[17] Bernard Montgomery was a confirmed teetotaler who intensely despised both liquor and tobacco. Unfortunately for the ruse, Meyrick James liked to smoke and was extremely fond of alcohol. Despite the CSM's desperate efforts to keep their star performer in line they were unsuccessful; the situation deteriorated, and astounding rumors soon reached SHAEF Headquarters in London that "Montgomery" had been seen weaving tipsily around the streets of Algiers while smoking a large, pungent cigar. To anyone familiar with the austere and abstemious commander, nothing could be more incredible.

The rumors from Algiers may or may not have been well founded, but the fact remains that Lieutenant James was soon packed off to England, Copperhead was shelved, and the whole matter was quietly abandoned. In his own account after the war, Monty's stand-in presented a somewhat glamorized and exaggerated version of the strange episode; however, researchers generally agree that the impersonation had little apparent impact on the Germans or their intelligence network. Actually no harm was caused by the CSM's abortive scheme, but it must be recorded that Meyrick James's most exciting play was forced to close after an unfortunately short run.

Montgomery in his own right, as well as Eisenhower and other luminaries, also made sham appearances in support of various Allied deceptions, particularly Skye and Quicksilver. And George Patton playing himself—a role in which he was unsurpassed—gave speeches and made numerous "inspection tours" as part of the FUSAG hoax aimed at the Pas de Calais. These impostures by key Allied leaders all gave added dimension and credence to the relevant cover operations.

Another notable "impersonation" involved the august Winston Churchill. During the war all of the prime minister's major trips were kept as secret and secure as possible, and he often traveled incognito. Although the XX Committee effectively controlled every German spy *in Britain,* there were numerous agents active in neutral countries, and any unguarded references or careless slips of the tongue could have had severe consequences.

In May 1943, Churchill was due to leave for the United States to meet with Roosevelt—the conference, incidentally, at which he received the good news about Operation Mincemeat. Plans had been made for the prime minister to sail on the *Queen Mary* which was then berthed in Scotland's River Clyde, and preparations for the journey were extensive. The main deck of the ship was sealed off from the rest of the passengers, and in this restricted area—also on the quayside—signs appeared in the Dutch language. A crack squad of Dutch soldiers drilled on the dock. Young liaison officers wearing the natty uniform of the Netherlands Army moved briskly between ship and shore. The area was soon buzzing with rumors that Holland's Queen Wilhelmina and her suite were traveling to America as guests of the U.S. government.[18] Cover and security were so effective that even some members of Churchill's War Cabinet, who were attending a separate conference at Hot Springs, were astounded when he appeared on board.

Churchill's daughter Mary, an officer in the A.T.S., was

traveling with him as his personal assistant; and she too had no inkling as to their destination. General Leslie Hollis, one of the Churchill's aides, reports that the prime minister informed his daughter at the dock where they were heading, then added with a grin, "I'm supposed to be Queen Wilhelmina."

Miss Churchill's eyebrows rose. "If only you'd told me," she replied promptly, "we'd have made you up to *look* like her."

Still another and far more sinister masquerade took place in December 1944 during the desperate German counteroffensive in the Ardennes area, known as the "Battle of the Bulge."

Preparations for that powerful but ultimately doomed assault were carried out in absolute total secrecy, and the plan was aided by bad weather which grounded Allied recon planes. Weeks earlier, false rumors had been spread by German agents implying that the Wehrmacht was near collapse, had used the last of its strategic reserves, and was now incapable of mounting a counterattack. All forward troop movements—from Monschau south to Echternach—were carried out at night, and during daylight hours the enemy front appeared quiet. To deaden the sound of their tanks and half-tracks moving up into position, the Germans laid beds of thick straw on their newly built log roads. To further cut truck noises, much of the ammunition for the opening barrage was carried up by hand, one round at a time.

John Toland, who recorded the history of the "Bulge," noted:

> Everywhere security was maximum. Radio silence prevailed, civilian phone service behind the lines was monitored. A camouflage officer was assigned to each village. A small army of special police roamed about, halting all unessential movement. . . . The troops were even issued charcoal, lest the smoke of wood-burning fires reveal their presence.[19]

As part of this last-ditch attempt by Hitler to turn the tide of war in the west, a special team of Wehrmacht commandos had undergone secret rigorous training. These men, under the direction of SS Colonel Otto Skorzeny, all spoke fluent English and were to be dressed in authentic American Army uniforms; infiltrated behind the lines of Omar Bradley's Twelfth U.S. Army Group, they would pose as American GIs and cause as much havoc and disruption as possible.

The hoax proved successful. Hitler's surprise Ardennes offensive—protected against air attack by fog, drizzle, and haze—exploded on December 17, and for a while made substantial and startling progress. Behind the lines, seven jeep-loads of Skorzeny's commandos, impersonating American soldiers, misdirected Bradley's regiments, turned signposts to point the wrong way, and tore up miles of telephone cable. Although there was only a small company of these impostors, they spread an enormous amount of confusion and alarm, and panicky rumors soon multiplied their numbers a thousandfold.

One rumor reached all the way to SHAEF Headquarters, then located at Versailles. According to this report, hundreds of Skorzeny's men, carefully disguised, had been parachuted near Paris and would attempt to kidnap or kill General Eisenhower. American security officers, believing the fabrication, quadrupled the guard at SHAEF, surrounded the compound with barbed wire, and brought in tanks and armored cars. For a while the Allied commander in chief—over his vehement objections—was kept virtually a prisoner.

Unknown to Eisenhower, he was also provided with a look-alike. One SHAEF officer, Colonel Baldwin B. Smith, bore a remarkable resemblance to the supreme commander, so he was dressed in one of Eisenhower's uniforms and periodically driven about the Versailles area to act as bait for the lurking paratroopers.

In actuality, there were no enemy forces within one hundred and fifty miles of Versailles, nor was Eisenhower in the slightest real danger at any time. Hitler's unrealistic counterattack was finally blunted and then smashed by stubborn Allied resistance; and by January 23 the battle was over and the Wehrmacht had been driven back to the line from which it had started. Many of Skorzeny's masqueraders, who had spread so much chaos, managed to return safely to their own units; the rest were killed or taken prisoner along with tens of thousands of their fellow Germans.

Some of these subterfuges and impersonations were far more crucial than others, but they each had their military value. Yet all pale into insignificance when measured against the Mincemeat deception which, in the history of *ruses de guerre*, stands alone in its simplicity and effectiveness.

Discussing the Mincemeat hoax Lord Ismay wrote, "The operation succeeded beyond our wildest dreams. To have spread-eagled the German defensive effort right across Europe, even to the extent of sending German vessels away from Sicily itself, was a remarkable achievement."[20] And Eisenhower's subsequent report that the defenders had been "badly deceived" was undeniably the result of the faked letters.

The counterfeit major's tombstone at Huelva carries a prosaic inscription, but a far better epitaph lies in the words with which Ewen Montagu introduced his own account of the episode. Montagu pointed out that the anonymous young man who died in England, abandoned and uncared-for, was able in his final incarnation to render a tremendous service—one which misled the enemy and in the process, ". . . saved many hundreds of British and American lives."[21]

Planting doctored maps, reports, letters, and battle plans was far from a new idea, but in Mincemeat it had been carried out with particular skill and perception. In that operation phantasm became hero; and between Ewen Montagu the composer and William Martin the masterwork a curious bond was evoked, prompting the Lieutenant Commander to say of his brilliant ectype, "In life he had done little for his country; but in death he did more than most could achieve in a lifetime of service."[22]

"Martin" and Sicily opened the door to Italy and signaled the start of an Allied push into continental Europe which, combined with the great Soviet drives in the east, would eventually smother the Third Reich.

But during all this, another campaign was raging halfway around the world. Here the battlefield consisted of overgrown jungle and parched volcanic terrain; and the defending enemy was, in many respects, more crafty and elusive than any the Allies faced in Europe or Africa.

14

The Conjury of Fanaticism

1942–1945

On Makin Atoll the combat patrol had been pinned down by a Japanese sniper hidden in a tangle of nearby ficus palms. Bullets from his 7.7 mm. automatic rifle splintered the coral rock just above the shallow draw where the GIs had taken cover. One of the American scouts, cradling his M-1, wormed his way slowly to the side, and after crawling some fifty or sixty yards raised his head cautiously. Now he could see the enemy rifleman silhouetted through the foliage, crouching warily on a high limb. The GI aimed carefully, squeezed the trigger, and the Japanese sniper, his arms flailing, fell to the ground. As the scout raised himself on one knee to signal to the rest of the squad a burst of 7.7 mm. fire caught him in the chest, bringing sudden death.

Later when the patrol succeeded in flushing out the real sniper, they found the scout's target. The dummy figure was made of clay and palm leaves dressed in an old jacket and peaked cap, with a wooden pole added to simulate a weapon.

Combat ruses in all their variations were used repeatedly by the enemy during the fighting in the Pacific. The Japanese became so wily and adept at guerrilla warfare and jungle illusions that the first American troops, baffled and confused, were almost led absurdly to believe that their opponents had somehow discovered the secret of invisibility. In actuality, geography in that theater of the war lent itself particularly well to camouflage and deception, and no troops anywhere made more brilliant use of *individual* cover and concealment than did the Imperial armies of Japan.

The Japanese, in the weeks following Pearl Harbor, had launched a crushing Wehrmacht-style *blitzkrieg* which had swept everything before it. Singapore . . . Hong Kong . . . Wake Island . . . Guam . . . the Gilberts . . . Borneo . . . Burma . . . northern New Guinea . . . Malaya . . . the East Indies . . . the Philippines . . . all crashed like tenpins; and within six months of their attack on Pearl the Japanese were in control of a Pacific empire bulging from Kiska and Attu in the Aleutians, south to the tip of New Guinea and sweeping around to Arakan in western Burma, as well as Korea, the island of Formosa and thousands of square miles of mainland China. This huge area contained almost four hundred million people, plus countless tons of vital strategic resources including rice, oil, rubber, tungsten, tin, chrome and manganese; and for a brief while the Japanese seemed indeed invincible.

Then came the Battle of Midway, a key encounter in which Admiral Chester Nimitz's U.S. carriers and cruisers delivered a crippling blow to the Japanese fleet. From that point on the tide in the Pacific was slowly reversed, and American, British, Australian, and Dutch troops, backed by the great industrial power of the United States, began the long return journey. Roosevelt, Marshall, and the other U.S. leaders, outraged by the Oahu disaster and its aftermath, were determined to right the balance swiftly and forcefully, and these goals had the country's total support.

"Not only had the Japanese misunderstood the temper of the American people when they bombed Pearl Harbor," wrote historian Abraham Rothberg, "they had also underestimated how rapidly America's productive capacity could replace the losses sustained there."[1]

Now, after their heavy reverses, the Allies began pushing in great, complex amphibious leaps from island to island, to defeat the enemy and shrink the Japanese perimeter. In the fall of 1942, under overall command of General MacArthur and Admiral Nimitz, American and Australian troops secured Buna in New Guinea after bloody fighting, and U.S. Marines captured Guadalcanal. Early the following year, American troops supported by air and naval units took the Russell Islands and began a reoccupation of the Aleutians. Next came New Georgia and Bougainville in the Solomons, and a continued drive along the treacherous northern coast of New Guinea, shaping a pincer movement which would slowly close on the Japanese mainland. Names and places that had once been totally unknown to the American public quickly became familiar as military communiqués and news reports were filed from Truk and Rabaul, Tarawa and Hansa Bay, Eniwetok and Kwajalein Atoll, Luzon and Mindanao, Saipan, Tinian, Iwo Jima, Okinawa, and a unique engineering miracle known as the Burma Road.

Wherever the Allies set foot the enemy fought back fanatically. The Japanese—veterans of a long war with China which had begun in 1937—had gained much combat experience, and had also won sufficient time to fortify their strongholds; and the areas they protected were in most cases made to order for shrewd military defense. Nowhere was terrain more suited to effective camouflage than in the Far East and the southwest Pacific, with its lush and varied vegetation. As the Americans and their Allies pushed forward, they had to battle not only the Japanese but tropical rain forests choked with tangled lianas and banyan roots; savannahs covered with kunai grass four to seven feet high; impassable sago and mangrove swamps; dense thickets of bamboo; and coastal strand forests teeming with casuarina shrubs, rubber trees, and nipa and coconut palms. Other areas consisted of open sun-blistered stretches of sand, volcanic ash and coral in which the enemy managed to dig numerous traps and tunnels.

A British Intelligence survey issued during the war reported:

> Very few examples of structural artificial camouflage are to be seen in the southwest Pacific area. The natural camouflage afforded by jungle and terrain in this theater is so abundant that the enemy has had to resort to little artificial construction.[2]

The Japanese took full advantage of this camoufleur's paradise, and used it to create hidden forts with deadly channels of enfilading fire. Gun emplacements were dug right into the massive root structures of banyan trees. Other strongpoints were buried in tangles of coconut logs covered with dirt, moss, and palm fronds. On the coral atolls, vast labyrinths with baffling twists and turns were carved through the sand and rock.

In disguising their khaki field uniforms, the Japanese also showed ingenuity, and instead of relying on patterned clothing made skillful use of natural materials. Foot soldiers in some areas were issued individual camouflage nets textured with bits and pieces of raffia fiber. These small nets could be tied together to conceal artillery, trucks, and tents. Other troops wore strawlike capes that covered their bodies, but could be quickly unlaced and thrown aside. Some Japanese caps were made of plaited rushes, daubed with green paint or mud to blend naturally with the surroundings. On Kiska and Attu, above the snow line, the patrols wore white parkas; and when moving across the open tundra they threw strips of grass matting over their shoulders. Many Japanese uniforms had a variety of loops and pockets sewn on the reverse side; when the jackets were turned inside-out the loops could be stuffed with branches and bits of local shrubbery, making the wearers a movable part of the underbrush.

Coupled with this ingenious personal concealment, the

Japanese outdid all combatants in the use of battlefield hoaxes. One marine sergeant recalled later, "There wasn't a GI in the Pacific who didn't have his own story of some nasty trick or other tried on him by the enemy."

In most arenas of the war, camouflage and deception were generally handled with panache and retained an almost chivalric quality; for all their deadly meaning, the military ruses employed in Europe and North Africa were a kind of intellectual game in which the goal was not to slaughter the enemy but to outwit him, and in the process to possibly minimize the bloodshed. In contrast, many Japanese combat ruses were brutal and cruel, and like their suicidal banzai charges, were frequently carried out with total disregard for the safety of the perpetrators. This ruthlessness, bordering at times on irrationality, stemmed from a cultural tradition largely alien to Western minds which placed national honor far above personal need. Their military code of rigid obedience and self-sacrifice made the Japanese unusually dangerous and cagey fighters, and added to the ordeals of Allied troops in the Pacific.

Enemy hoaxes were lethal and varied. In addition to dummy snipers, the Japanese at different times and places—

. . . dragged the bodies of dead or wounded American soldiers close to their own lines, and propped them up to lure rescue attempts.

. . . scattered cast-off garments and equipment along a jungle trail to create an illusion of hasty retreat, then waited in hiding for their pursuers.

. . . set up decoy machine-gun posts in obvious locations, covered by camouflaged snipers.

. . . fired captured American weapons to give the impression that a position was in friendly hands, drawing patrols into ambush.

. . . dug heavily concealed foxholes with aiming slits only at the back, enabling them to cut down advancing GIs from the rear.

. . . synchronized their mortar and artillery barrages with

American mortars, to create an impression that the Americans were accidentally shelling their own units.

In Malaya and New Guinea, enemy troops wore cutout cardboard rings over their peaked caps to simulate the wide-brimmed hats of the Australians. As they slipped from tree to tree through the dense forests, this simple device created the expected confusion.

The Japanese attempted vocal ruses, particularly at night; and MacArthur's weary riflemen, crouched in their foxholes on Leyte and Tulagi, Saipan and Guam, soon grew used to shrill, high-pitched cries drifting through the blackness. *"Hey, Joe! Come here!" "Doc, I am wounded! Doc, Doc! Give me aid!" "Buddy, help me! I am bad hurt!"* Before attacking they also screamed taunts and curses, not only to boost their own morale but to rattle the GIs, draw answering fire, and pinpoint the American positions.

In some cases the tenacious Japanese used their bodies as decoys, feigning death until American patrols came close enough to be attacked. Others deliberately showed themselves in order to attract gunfire and reveal the exact whereabouts of their enemy.

During the island campaigns the GIs found many instances of snipers tied into trees; this would help to steady them in case they were wounded, but it had a more macabre function as well. A 1943 U.S. combat intelligence report quoted a marine officer on Guadalcanal: "I believe that one reason the Japanese ordered their snipers to tie themselves in trees was to get us to waste our ammunition. When a sniper tied to a tree is killed, he doesn't fall. As other soldiers pass by later, they again spray the body with bullets. I cut down the body of one Jap who had been dead at least three days. I counted 78 bullet holes."[3]

In the hands of this obsessive and determined enemy, the ground itself became a weapon. Tunnel-digging has always been a valuable defense tactic, and in World War II it served its purpose in strongholds such as the Maginot and Siegfried Lines, Gibraltar, Corregidor in the Philippines, and in the Harz Mountains of central Germany; but the camouflage-conscious Japanese in the Pacific developed tunneling to a high military art.

Wading ashore on Tarawa in November of 1943, U.S. Marines found themselves facing a tunnel network of two hundred interlocked bunkers and coconut log barricades and each, Hanson Baldwin noted in his account of the battle, had to be approached as "a separate problem in assault."[4] In addition, ". . . snipers are everywhere, lashed to the riddled palms, behind a hummock of sand, under the pier." The Japanese fought almost to the last man before the island was taken.

Early in 1944 came Kwajalein and Eniwetok in the Marshall Islands. Prior to the assault on Eniwetok, U.S. battleships, cruisers, and destroyers had shelled the base heavily, and Navy Avengers followed up with low-level bombing and strafing attacks. Author-historian Merle Miller, then an army sergeant, wrote in *Yank* magazine that during the barrage, "all of Eniwetok seemed on fire."[5] But following the shattering bombardment there were no enemy responses. "None of our observers," Miller recalled, "had sighted a single Jap on the island—or any other living thing. Some of the men of the 106th Infantry wondered out loud whether . . . the Japs had fled without a fight."

Only after the fourth wave of U.S. troops had hit the beach and moved inland did the guerrillas begin to show themselves, firing rifles and knee-mortars from hidden positions and taking the GIs by surprise. The Japanese had literally buried themselves in small, well-protected foxholes, and allowed whole battalions to move past them before attacking the Americans from the rear. Some of these hidden burrows were within *a few yards* of the advancing platoons; and at one point, Miller noted, ". . . there were nearly as many Japs behind as in front of our own lines."

Similar mole camouflage awaited the Allies in the Palaus

and also the Philippines, but of all the skillfully built strongpoints, dugouts, command posts, and subterranean trench networks, none were more effective than those on the island of Iwo Jima which, Abraham Rothberg wrote, ". . . was a maze of reinforced concrete pillboxes, bunkers and blockhouses, beautifully camouflaged and connected by a labyrinth of underground tunnels and corridors."[6]

Iwo, a small volcanic island in the Bonins, was the best available air base between Saipan and Tokyo, and the Japanese had been using it as an advanced radar and fighter station from which their planes intercepted B-29 Superfortresses bound for Kyushu and Honshu. The Allies needed the island as a base for P-51 Mustangs that could act as fighter escorts for these powerful bomber formations now beginning to strike at airfields, factories, and other targets on Japan's home islands. Equally important, Iwo would serve as an emergency landing field for B-29s which had been crippled by enemy flak, and couldn't possibly get back safely to their bases on Saipan.

In February 1945, after two months of shelling and bombing by the U.S. Fifth Fleet to neutralize enemy defenses, three marine divisions stormed ashore on Iwo Jima. In spite of the heavy preliminary bombardment, twenty-two thousand defenders—operating from their tunneled warrens—fought back fiercely. Time and again the marines, after clearing a combat area, found that General Kurabayashi's troops had infiltrated through their mazes and reappeared miles behind the front to begin fighting all over again. One of the toughest battles was for control of Mount Suribachi, a 546-foot high extinct volcano at the southern end of the island honeycombed with defenses. Suribachi changed hands five times before the marines finally secured it, and the raising of the American flag at its peak produced a dramatic photograph which gained wide fame and is now recreated in bronze on the bank of Washington's Potomac River.

Former Staff Sergeant Irwin B. Nelson, who served with a Twentieth AF Fighter Squadron on Iwo Jima, recalled the tactics of the Japanese. "During the barrage our ships and planes didn't do any pinpoint bombing; Iwo was so small that the entire island became a target. But the Japs lived through it by just slipping underground—and when the marines came in, they were there waiting."[7]

For almost a year Nelson operated a UHF homing station on Suribachi to guide U.S. fighter planes escorting the giant Superforts. "Months after we thought the island was finally pacified," the ex-airman added, "the Japanese were still coming out of tunnels ready to fight, even with their situation completely hopeless."

An incident recounted by Irwin Nelson perhaps typifies this cagey and unremitting behavior. "Alongside our airstrip," he said, "there was a big pyramid of empty fifty-gallon oil drums. They were being used for holding sand, gravel, oily rags—all the usual odds and ends of airfield maintenance. One night a Japanese sniper with a rifle, grenades, and some food sneaked down and hid in one of the empty drums. For camouflage he pulled a piece of burlap over himself. The sniper actually lived there in the oil drum, right under our noses, waiting for his chance; but fortunately he was caught before he did any damage."

In Pacific combat, guile and deception were as much a part of life as mud, sun, torrential rain, and snapping ants; but in areas away from the front lines the enemy ruses were inconsistent, showing versatility in some cases and ineptitude in others. During their earlier conquests the Japanese, like the Nazis, had little need for defensive camouflage. Then in mid-1943, with the Allies taking the initiative and gaining air supremacy, the defenders began industriously creating false targets to draw off their attackers.

All through the South Pacific, from the Malay Peninsula east to the Bismark Sea, scores of Japanese field decoys began to appear on photos of the U.S. Fourth and Seventeenth Aerial Reconnaissance Groups. At Buna, on the northeast coast of

New Guinea, many fake howitzers made of coconut logs were installed. Decoy artillery—some of it fairly crude—was also displayed in places such as Lae, Tobera, Malahang, Boram, Vila and Kolambangara. At Rabaul, a major enemy base on the island of New Britain, military supplies were stacked and covered with thatching, to resemble native huts and farm buildings. And at Cape Gloucester, also on New Britain, the Japanese simulated a four-gun heavy AA battery serviced by a "crew" of eighteen dummy figures made of dried grass stuffed into surplus uniforms. Many of these frauds were easy to detect, but the Cape Gloucester illusion proved so convincing that it was attacked numerous times by Allied aircraft.

The Japanese also took pains to conceal their precious squadrons of Mitsubishis, Nakajimas, and Nagoya Zeros, as well as their airfields; and some of their ruses, while they lacked the subtle finesse of the British during the Luftwaffe raids of 1940-41, were inventive and original.

On the island of New Georgia in the Solomons are thick stands of palm trees, and the Japanese intended to build a fighter airstrip in one of these groves. But by that time Allied long-range photo recon planes (converted B-17s and stripped-down Liberator bombers) were coming over in increasing numbers. The problem for the Japanese camoufleurs was not only to hide the new airstrip but to keep its construction a secret, and their solution was a unique one. Using old wire and telephone cable, they first tied the tops of the palm trees securely in place. Next they sawed off the trees just below this point, then removed the trunks and cleared their runway area. The tops of the palms, wired into position, created an excellent natural canopy that concealed the work going on underneath. Later the Japanese fighter pilots were given key landmarks and were able to find their way safely to the covered airstrip. Since there was little or no change in the landscape when seen from above, the field remained undetected until reported to the Allies by natives hiding in the island's jungles.

At other bases, aircraft were placed under bamboo frames covered with rattan, or hidden under masses of palm fronds. The resourceful Japanese camouflaged *wrecked* planes as well as operational ones, making it impossible for Allied photo interpreters or attacking pilots to single out the legitimate targets. On Wake Island, before it was reoccupied, the Japanese dug underground hangars and carefully covered them with earth and foliage. At Kavieng, on the northwest tip of New Ireland, an airstrip was surfaced with crushed coral, then painted to simulate the plowing patterns on adjoining fields. Short of paint, the camoufleurs used drained fuel oil to create their aerial illusion of plowed farmland.

A device especially favored by the Japanese was to display damaged and unfit aircraft along the edges of their runways where they could—and did—attract the attention of American fighter-bombers. At one large base in southern Burma, the camoufleurs also used a nearby airplane graveyard as a hiding place. Each fighter plane, as it finished its mission, taxied to the edge of this aircraft junkyard and was guided to a prearranged spot in the wreckage. The effort involved in the hoax was more than offset by the success of the concealment.

Later, as the war intensified and Allied bombings continued to drain Japanese air strength, the commanders in desperation fell back on that most dependable of all defensive ruses: the use of decoys and dummies. Imitation Zeros and Mitsubishi bombers made of canvas and bamboo, painted green with the characteristic red circles on their wings, were built and installed everywhere. Hundreds of these shams were created for use at forward airstrips and bases deep inside Japan itself; and in 1946, the U.S. Strategic Bombing Survey for the Pacific area estimated that decoys and dummies made up as much as *twenty-five percent* of all aircraft visible on Japanese fields.[8]

These, like Turner's numerous decoys in England, inevitably drew attacks away from legitimate air squadrons that the Japanese were now husbanding for what they referred to as "The Final Battle on the Homeland."

In camouflaging their factories and other industrial targets, the Japanese were far less successful. During November 1944, Boeing B-29s of General Curtis LeMay's U.S. Twentieth Air Force had begun regular mass raids on objectives in major enemy cities; the huge bombers, based on Saipan in the Marianas, poured thousands of tons of incendiaries and high explosives on Tokyo, Japan's main industrial and military center, as well as key areas such as Osaka, Nagoya and Kobe. Enemy oil refineries, rail yards, aircraft plants, supply depots, and steel and munitions factories were hit repeatedly. By mid-March 1945, Japanese production had been badly crippled and some sixteen square miles in the once-productive core of Tokyo itself had been completely burned out.

Like the British and Germans—and Americans on the West Coast after Pearl Harbor—the Japanese tried to protect certain targets with paint and camouflage; but they apparently misunderstood the basic purpose of disruptive painting and concealment. Instead of using color and pattern to tone down these installations and blend them with their surroundings, the camofleurs did just the opposite: Many crucial structures were naively "dazzle-painted" in highly visible designs unrelated to the landscape. Oil tanks at a depot near Kure were given a coat of bold rectangles and diagonal slashes. Hangars at the Yokosuka seaplane base were dazzle-painted in large, vivid, geometric forms. Similar unlikely "op art" was applied to factory buildings in Kobe and Osaka. Even some railroad trains were splashed with erratic blobs and patterns of bright color, which only served to increase their visibility.

Unequaled at individual concealment and jungle camouflage, the Japanese somehow failed to visualize how larger installations would and should look from the air, and as one B-29 navigator put it, ". . . they made our work easier for us."

In the final months the Japanese had begun to correct these mistakes and also to realize that tunneling, effective on their island outposts, could be used for industrial cover as well. Emulating the Germans who had created vast tunnels for their V-2 secret weapon factories, the Japanese administrators began putting their own military production below ground. The 1946 U.S. Bombing Survey revealed: "By the end of the war, although only a few underground plants were actually in operation, the construction of tunnel-type factories . . . was beginning to assume major importance."[9]

If the conflict had gone on a while longer, the Survey added, "an important percentage of Japan's essential war plants might have been operating underground."

One of these newly completed installations was a huge naval arsenal and aircraft factory located in the hills of Yokosuka, at the southern end of Tokyo Bay. This station contained some three hundred fifty thousand square feet of underground space, housing laboratories, machine shops, storage areas, power units, offices, even aircraft hangars, all linked in one vast network. Photo interpreters, studying aerial prints of the Tokyo Bay area, had readily pinpointed the tunnel entrances and traced their access roads; however, a later PI evaluation stated: "The extent and function of the different tunnels were not realized, nor was it understood that all were connected to form a single system."[10]

The Yokosuka tunnel complex would doubtless have been replicated in other areas, but time was running out for Japan; and except for the *Gumbatsu*, the military diehards of whom there were an alarming number anxious to fight to the last survivor, the Japanese Empire and its "Greater East Asia Co-Prosperity Sphere" were approaching collapse.

In the year 1281, according to Japanese history and legend, a powerful Mongolian war fleet was en route to attack the islands. Off the coast of Japan the invading armada was struck by a sudden tornado and destroyed. Tenshi, the Son of Heaven, had intervened by sending the Divine Wind—the "kamikaze"—to protect the Japanese homeland.

Now, centuries later, Vice Admiral Takijiro Onishi's

Special Attack Force, the "Knights of the Divine Wind," set out on a desperate and futile mission to repeat the miracle and break the overwhelming power of an invader. Young Samurai-obsessed volunteer pilots, flying obsolete, stripped-down aircraft packed with explosives, had been trained to deliberately crash their planes into American vessels, in a lethal exchange of "one plane for one ship." Their sacrificial kamikaze attacks reached a crescendo in April and May of 1945, during the final great battle for the island of Okinawa in the Ryukyus, and they did appalling damage. In all, the Japanese launched an estimated twenty-five hundred suicide sorties, scored four hundred seventy-five direct hits, sank forty American fighting ships, and seriously crippled hundreds of others.

A survey of World War II Pacific camouflage must include mention of one vital but little-known aspect of these suicidal missions. Many kamikaze planes were brought down by accurate U.S. antiaircraft fire as the pilots dove toward their targets, some were demolished in air-to-air fighting, and others were destroyed on the ground. But Onishi's lethal squadrons kept on coming; and the seemingly endless flow was made possible in part by the superb way in which those particular planes had been concealed until ready to be used. From the very first appearance of the kamikazes, during operations at Leyte and Lingayen Gulfs, an Allied visual and photographic hunt for the aircraft had gone on relentlessly; but the results had been generally disappointing. The bulk of the suicide squadrons were known to be deployed at small heavily camouflaged bases on the island of Kyushu, and from there more than fifteen hundred missions were launched. Yet unknown to the Allied searchers, scores of suicide planes were hidden on Formosa as well; they had been widely dispersed in thickly wooded areas, and some were literally dismantled and put under cover in Formosan towns and villages.

Photo recce missions were flown repeatedly over these areas, but the concealment was so successful that it was never effectively penetrated. While Army Intelligence units estimated that eighty or ninety kamikazes might be based on Formosa, there were in fact seven hundred planes hidden there, and many were dispatched on the one-way death missions over Okinawa.[11]

These last-gasp enemy efforts, and similar measures, were dwarfed by the appalling mushroom clouds that rose in August 1945 over Hiroshima and Nagasaki.

Atomic destruction brought an end to Japanese resistance, making unnecessary a frontal assault on the home islands with its incalculable suffering, bloodshed, and death. On September 2 came the unconditional surrender of the most ingenious, most tenacious, and certainly most ruthless of the Axis powers.

15

Navies in Motley

1940–1945

Scholarly, donnish John Graham Kerr was neither an artist nor a sailor; on the contrary, he had by 1914 become one of England's leading scientists whose research work in zoology and embryology had won much recognition. Then the Great War exploded, sending its shock waves into the most cloistered of laboratories and shaking John Kerr's detachment with surprising military results.

In the years leading to Sarajevo, Germany had been feverishly building her navy in an effort to equal and eventually overtake Britain's maritime supremacy; and now with war a reality, the Kaiser's fleet under Admiral Alfred von Tirpitz had become a formidable one. British shipping in the early 1900s accounted for over forty percent of the world's total merchant tonnage, and two-thirds of England's own food had to be imported, which made the country especially vulnerable to Tirpitz's heavy cruisers, destroyers, armed steamships, and above all his deadly U-boats. "Fear of the unknown but certainly bellicose intentions of the enemy," Barbara Tuchman noted, "and particularly fear of the invisible submarine, whose lethal potential loomed more alarmingly each year, made for a highly sensitive state of British naval nerves."[1]

Erudite John Kerr, brooding like millions of Englishmen over this German threat, wondered what if anything he might possibly do.[2] Embryologists in wartime were hardly in great demand, but studying the multihued, oddly patterned marine vertebrates on his laboratory shelves, he chanced on a theory that took him hurriedly to the offices of the British Admiralty. The forty-five-year-old scientist, too distinguished to be brushed aside as a crank, succeeded in impressing the naval commanders with his unique idea for disguising ships; and before long British seagoing vessels began to appear with gaudy and outlandish patterns decorating their hulls. So Kerr found himself the somewhat unlikely father of the "dazzle" principle of ship camouflage, a bizarre visual signature identified with the naval actions of both world wars.

Dazzle camouflage was "fool the eye" applied to the nautical dimension. As conceived by Kerr, it involved painting a ship so as to create a false perspective, and to distort the vessel's true course and speed. This in theory would mislead officers aboard enemy submarines who sighted through calibrated periscopes before launching their torpedoes. Since the range officer had to "track" a moving ship, the distortion would make aiming difficult; and might cause a torpedo to miss its mark, or hit a less vulnerable area.

The actual value, if any, of this type of seaborne deception has been widely debated, but the attitudes of the British and U.S. Navies in both wars was that such designs did no harm, could conceivably cause trouble for the enemy, and were a definite help to morale, particularly among merchant seamen who tended to feel more comfortable on camouflaged vessels.

During World War II the coming of radar, sonar, and other electronic aiming devices minimized the need for visual rangefinding and made dazzle camouflage largely redundant. But the practice was nonetheless continued, and Allied ships in painted motley sailed all the sea routes of the world, from Murmansk to the Drake Passage, from Puget Sound to Darwin, from the Cape of Good Hope through the Indian Ocean to ports on the Bay of Bengal.

Dazzle design for England's naval and merchant fleets came under the Research and Development Section of the Admiralty, and the original theories proposed by Kerr were later expanded by the noted artist and naturalist, Peter Scott. In the United States, ship camouflage—initially handled by the Navy's Bureau of Construction and Repair—was supervised during World War I by distinguished American artist Abbott H. Thayer, whose paintings now hang in the Boston Museum of Fine Arts, Washington's Corcoran Gallery, and the Metropolitan Museum of Art in New York City. Later this work came under the Camouflage Section of the U.S. Navy's huge experimental base at Anacostia, on the Potomac River.

Here at Anacostia in World War II, designers and draftsmen, working under the leadership of artist Everett Warner, experimented with miniature ship models to create hundreds of unusual patterns and optical effects. These deceptions —with their familiar curves, whirls, stripes, and angular slashes—were then transferred to "Camouflage Design Sheets" showing detailed formats for port, starboard, bow, stern, and superstructure. The layouts also gave information as to correct paint tones, known as "measures," which were chiefly shades and hues of blue, gray, and green.[3]

Lieutenant SG Robert Paulus served aboard the U.S. merchant transport *Henry Gibbons*, which had been specifically dazzle-painted for the Murmansk run. "We were scheduled for the North Atlantic," Paulus said, "so the whole ship had been covered in wavy horizontal bands of white and battleship gray, to fit that area—especially when we came through the ice floes of the Barents Sea."[4] The *Henry Gibbons* was soon reassigned to transporting American troops to Naples for Mark Clark's Italian campaign, and the paintwork was suitably modified.

Recalling that period in the U.S. Merchant Marine, ex-officer Paulus added, "We seemed pretty casual about that type of camouflage, but the British were incredible. Their ships, especially the transports, carried every type of crazy pattern and color. When we sailed in convoy with them, it was like being in the middle of a floating art musuem."

Britain's maritime commanders—after their traumatic losses during the 1941–42 Battle of the Atlantic—generally paid more attention to dazzle camouflage than did their American allies. In the early phase of the war, merchant ships had two main passive defenses against submarine attacks: one was deceptive painting; the other was to steer an evasive or zigzag course. But many *active* antisub measures were soon developed, and proved far more effective. These included the use of escorted convoys, the dropping of powerful depth charges, sonar detection, aerial reconnaissance, and the heavy bombing of Germany's U-boat bases. As the U.S. Navy grew in strength, American strategy emphasized these active policies; and while some U.S. ships wore dazzle paint, many were simply given a compromise tone of haze gray ("Measure 14"), which reduced visibility at night and also during daylight hours.

In addition to course distortion, other optical tricks were attempted. Some combat vessels had large curling "waves" painted on their bows. These fake bow waves gave the impression that the ship was traveling at a greater speed than was actually the case. A number of ships were given false waterlines. The raised band was colored to blend with the surface of the sea, and when seen from a distance would theoretically cause submarines to overestimate the ship's range. Other *trompe l'oeil* experiments involved painting the dark silhouette of a destroyer on the hull of a larger vessel such as a heavy cruiser; or outlining the form of a small ship directly on the deck of an aircraft carrier, to confuse attacking planes. The Japanese carrier *Zuiho,* for example, displayed a painted version of a light cruiser—with ingenious three-dimensional shadows—on its broad flight deck. This didn't prevent Admiral Raymond Spruance's bombers from sinking the *Zuiho* during the great

Battle of the Philippine Sea in June 1944; and such odd illusions were generally found to be of little real value.

At times the German Navy also used paint camouflage, with some success. Months before the United States entered the war, Hitler's Kriegsmarine had been sending secret weather-reporting missions to the Greenland area. One German vessel, the *Buskoe*, had been carefully painted an overall dull white, and when lying in the shadows against one of the giant icebergs in that area, the little trawler became almost invisible.

In April 1941, to assist the British in the crucial Battle of the Atlantic, the United States had made Greenland an official protectorate. The *Buskoe*, ignoring this, prowled the northern sea-lanes, slipping silently from fjord to iceberg, establishing espionage and transmission stations on Greenland territory. In September of that year, after a long search, the elusive sea ghost was finally caught by the U.S. Coast Guard cutter *Northland*, which had been similarly camouflaged. The *Northland*'s crew found the captured ship loaded with antennae and high-frequency radio gear—one of the many attempts by the enemy to set up intelligence bases in that part of the world.

Later in the war the Germans used more conventional cover to protect their dwindling naval strength from Allied air power. The battle cruisers *Scharnhorst* and *Gneisenau*, moored at Brest, were draped with netting and canvas, not to conceal their presence which was obvious enough to Allied intelligence, but to obscure vulnerable points from raiding aircraft. Similar treatment was given Hitler's most powerful battleship, the 42,900-ton *Tirpitz*, which lurked menacingly in the steep protective fjords of Norway. Rafts covered with foliage were made fast to the ship's bow and stern, and camouflage nets were extended from them to her hull. Other nets were stretched over her superstructure. At Alten Fjord where the battleship came under frequent attack, her commander, Captain Hans Meyer, also made heavy use of defensive smoke screens. In spite of all these attempts, Allied bombers finally sent the *Tirpitz* to the bottom in November 1944.

The Kriegsmarine's strategists developed another, less passive naval ruse which—in concept if not in practice—showed inventiveness worthy of a Maskelyne. The hoax involved a "Water Donkey," or *Wasseresel,* which consisted of a full-scale decoy submarine conning tower. This simulation, packed with high explosives, was designed to be towed along the surface by a submerged U-boat at the end of a long cable. Allied warships would (so the Germans hoped) mistake the fake conning tower for a genuine submarine and attempt to ram it, setting off a deadly explosion. There is no record that a *Wasseresel* was ever successfully employed. Admiral Erich Raeder's U-boat commanders were understandably reluctant to put to sea with so lethal an encumbrance, and certainly didn't want to be delayed by a heavy tow line while being hunted by Allied destroyers. The *Wasseresels* were duly built, but according to one account, "Unofficial reports seem to show that most of the few which were issued were 'accidentally lost at sea' shortly after leaving harbor."[5]

In the long history of naval intrigue and deception, the Germans also wrote one chapter of special distinction—a chapter reminiscent of the more courtly and less complex age of wooden sailing frigates.[6] It began on May 3, 1940, as the British merchant vessel *Scientist* steamed along the west coast of Africa, bound for Liverpool from Durban with a cargo including 1,150 tons of chromium and 2,500 tons of maize. The *Scientist*'s captain, alert for enemy U-boats and surface ships, scanned the horizon anxiously. Nothing was in sight, however, except for an old rust-stained freighter lumbering slowly toward them flying the flag of Japan, a country which at that time was still neutral. Through his glasses the British skipper noted that it was the *Kasii Maru;* he could see a number of Japanese sailors with scarves around their heads and their shirttails out, lounging aimlessly at the rails. A woman—either a passenger or the wife of the Japanese captain—was pushing a baby carriage along the freighter's shabby deck.

Suddenly the startled British commander received a blunt signal to heave to and not use his wireless. Moments later the *Kasii Maru* hoisted the German flag and opened fire. The *Scientist* stopped and her crew hurriedly began to lower the boats; but before the vessel was abandoned and torpedoed the radio operator was able to signal QQQ, a cipher which meant "I am being attacked by a disguised merchant ship." The heavily armed German raider managed—as it would do on numerous occasions—to pick up all the survivors of the sunken ship before moving on.

This was the first attack on Allied shipping by the notorious *Atlantis,* one of a small handful of secret marauders sent out by the Kriegsmarine to harass Britain and interdict her maritime lifelines. The *Atlantis,* under command of an outstanding and scrupulous officer, Captain Bernhard Rogge, remained at sea continuously for twenty-one months. During that time she sank or captured twenty-two Allied freighters carrying a variety of important cargoes such as steel, chemicals, leather, machinery, teak, chromium, fuel and diesel oil, wheat, maize, textiles, and pig iron.

In the amazing course of her wanderings, Rogge's raider assumed more nautical disguises than a character actor playing a long season of Shakespearean repertory. Prowling the South Atlantic, the Indian Ocean, and the convoy lanes of the Dutch East Indies, the *Atlantis* at varying times masqueraded as, among others, a Russian ship, the *Krim;* the Japanese *Kasii Maru;* a Dutch steamer, the *Abbekerk;* and the British merchant-cruiser *Antenor.* Beneath these various and apparently harmless disguises was a small but deadly fighter. The versatile ship was powered by large diesel engines which gave her a great radius of movement. Stowed on board were 4,200 tons of oil and coal, 400 tons of food, and 1,200 tons of fresh water. Behind collapsible bulkheads along her decks, she carried six 5.9-inch guns, one 75 mm. gun and 6 light antiaircraft guns. In addition there were 4 torpedo tubes, dozens of mines, and a well-hidden Heinkel 114 seaplane, to be used for reconnaissance. Completing her disguise as an ordinary freighter, she had a fake stack and several dummy "cargo booms," and below decks was an assortment of foreign naval uniforms and other clothing, both male and female, which the crew wore at appropriate times. The raider also carried a large supply of paint that was used frequently to change her name and vary the colors of hull and superstructure.

According to the conventions of so-called civilized naval warfare, a captain may give his vessel any disguise or attempt any hoax he chooses, provided that *before* he opens an attack, he displays his ship's correct national flag. Bernhard Rogge followed this somewhat anachronistic code meticulously. He also made every effort to rescue the survivors of all sunken ships, and treated his prisoners with the greatest care and consideration.

Anxious to keep his activities covert for as long as possible, Rogge had trained his gunners to cripple a freighter's wireless apparatus with their very first salvos; and the rest of the operation could then proceed at a more leisurely pace. In spite of this, hasty QQQ signals from ships under fire soon began to reach the British Admiralty. As a pattern emerged, the Royal Navy tried vigorously to track down the elusive Rogge, tying up numerous ships and planes that were badly needed elsewhere. But the *Atlantis*'s astute skipper, rendezvousing periodically with German supply ships and restocking water at remote islands in the South Pacific, managed to elude every naval unit sent to hunt him down.

In November 1941, the raider was finally intercepted by the British cruiser *Devonshire* under Captain R. D. Oliver. The *Devonshire,* much faster and more heavily armed than the German vessel, opened fire at long range; and Rogge, aware that he was now outmatched and outgunned, tried one final ruse. He signaled urgently (and indignantly) that he was a friendly vessel, the British ship *Polyphemus.* He then gave the letters RRR, an Allied cipher indicating that an enemy warship was nearby. Unknown to Rogge, that particular signal had recently been changed from a cluster of three Rs to four Rs. This increased the *Devonshire* captain's suspicions, and after checking by radio he soon received word from Freetown in Sierra Leone that the mysterious ship couldn't possibly be the *Polyphemus,* which was known to be elsewhere. Once again Oliver directed a withering fire at the *Atlantis;* Rogge and his crew had to take to their boats, and the deadly little raider went down at last.

Captain Oliver had also been warned of the presence of enemy U-boats in the immediate area, so he made off without gathering up Rogge and the other survivors. They were later rescued by Axis submarines and eventually made their way safely to German territory.

In all, the Kriegsmarine operated nine raiders disguised like the *Atlantis* as innocent merchantmen, and chief among these were the *Orion, Pinguin,* and *Thor.* Between them, the German masqueraders sank or captured over 850,000 tons of shipping and kept Allied search units busy for three and a half years before the danger was completely eliminated. During World War I the British had also used armed, disguised freighters—known as "Q ships"—which had been moderately successful. And on Cyprus in 1942, the redoubtable Jasper Maskelyne camouflaged a number of powerful RAF speedboats as old Greek "fishing caiques," for use in undercover activities. But no other warships incognito ever matched the record of the *Atlantis,* which logged over 100,000 miles of sea travel during a remarkable war life of 622 days.

Her captain, skillful and humane, was later decorated, promoted to admiral, and given a high post in the German Navy. More significantly, he won the great admiration of many Allied commanders whose ships had been destroyed by his hidden guns.

In a different theater of war, other forms of naval disguise and deception took place. The southwest Pacific, with its lush profusion of foliage and myriad overgrown islands, lent itself readily to techniques of natural cover; and at sea, just as on land, the Americans and Japanese made use of these abundant materials, camouflaging their ships with bushes, shrubs, palm fronds, and nets festooned with moss and dried grass.

Most U.S. warships in that area wore coats of haze gray low-visibility paint, with occasional destroyers, cruisers, or auxiliary carriers showing elaborate Anacostia dazzle patterns. Small vessels such as PT boats, used for inshore operations, were generally painted forest green to help them blend with the verdant coastlines. When a PT boat was moored in a tropical cove or inlet, a handful of sago or nipa branches completed the necessary illusion. PT-109, made famous by its skipper, then Lieutenant John F. Kennedy, was regularly camouflaged in this manner.

The Japanese also used foliage cover, and some of their vulnerable cargo transports were so laden with palm fronds and assorted vegetation that, from the air, they could pass for uninhabited jungle islands. While these vessels rode at anchor the concealment was highly successful, but nobody has ever discovered how to camouflage the wake of a moving ship, and Allied submarines and aircraft sank many Japanese freighters on the open sea, palm fronds and all.

By mid-June 1945, the historic battle for Okinawa had ended and the key island was firmly in American hands. At that point Japan's once formidable fleet had been shattered, and only a few last vestiges remained. Admiral William "Bull" Halsey's carriers, wrote historian Edwin P. Hoyt,

> . . . struck the Japanese islands repeatedly against very little opposition, and Admiral [John F.] Shafroth struck with a bombardment force of battleships and cruisers. From now on until the end of the war, the Americans prowled the Japanese waters almost at will, exercising their overwhelming naval and air force.[7]

Although the enemy's power was in ruins the spirit of defiance was very much alive, and Japan's commanders prepared fanatically for a final cataclysmic battle on their home islands. But first this called for the protection of every bit of their remaining strength.

At Kure, on the coast of Honshu and the shore of the Inland Sea, there is an excellent, well-protected natural harbor. Here the Imperial Navy gathered the remnants of its battered fleet and concealed them in what the U.S. Bombing Survey called "the most extensive Japanese camouflage effort of any kind."[8] The ships—one cruiser, two light carriers, and a few destroyers—were literally dug into hills and coastlines in the Kure area. Nets were strung from shore to superstructures, then covered heavily with foliage. In some cases the Japanese transplanted whole palm trees from the countryside to large tubs on the decks of the sheltered vessels, and gaps in between were strewn with moss and underbrush. In effect, the ships became an extension of the Honshu landscape.

This remarkable use of defensive camouflage had, of course, one obvious drawback: It completely immobilized the concealed warships. Before long, despite their elaborate cover, the ships were spotted by skilled American PIs, and target photos were prepared for U.S. carrier task forces. With the Pacific war in its final weeks, the last of the Japanese Imperial Navy was smashed and sent to the bottom in the deadly Kure strikes of July 1945.

One unusual American hoax—widely documented in recent years—requires mention here as a superb and classic example of the use of a decoy; but this particular naval decoy was cryptographic rather than physical.

In May 1942, at the height of its aggressive, lightning conquests, the Japanese planned a major attack on Midway Island in the Central Pacific. Midway, Japan's springboard for a future invasion of Hawaii, was vital to the U.S. defense network and its loss would have been catastrophic. The enemy's huge assault fleet under Admiral Isoroku Yamamoto consisted of eleven battleships, five aircraft carriers, fourteen cruisers, seventeen submarines and over fifty destroyers, plus hundreds of planes. Some of these units were slated for a diversionary thrust at the Aleutians, but the vast bulk would strike directly at Midway. Opposing them, the American fleet could muster only three carriers, eight cruisers, seventeen destroyers, and twenty-five submarines. Heavily outmatched, the best hope for Admiral Nimitz and his commanders was to keep their small force intact and *in the right spot,* and not to be lured out of position or caught off balance by feints and diversions. Considering the huge stretches of open Pacific, extending over thousands of miles, Nimitz's chances were extremely slender.

But despite these lopsided odds, the defenders had a hidden weapon that would prove of incalculable help. Unknown to the enemy, American cryptanalysts, after months of painstaking work, had succeeded in cracking the secret Japanese code. This operation, known as "Magic," paralleled the remarkable Ultra coup in the European theater which had broken the code of the Wehrmacht. With the help of their Magic intercepts, the Pacific cryptologists had learned a great deal about Yamamoto's plans and fleet dispositions; but Allied intelligence still lacked details as to the precise date, time, and place of his coming attack. Magic indicated that the letters AF (intercepted several weeks earlier) stood for the locus of the main Japanese assault. By a process of elimination, this codegroup pointed to Midway—but there were other possibilities. What did AF really mean? When and especially *where* would the Japanese strike? In his account of this dilemma, code expert David Kahn wrote:

> There was no clear-cut answer. Several Japanese strategies appeared possible. Nimitz himself thought Midway was the target, but in Washington Admiral Ernest J. King, Chief of Naval Operations, who was working from essentially the same information, concluded that Oahu was.[9]

In Washington the tension increased, and so did the pressure for an accurate, *positive* identification of AF. The stakes were extremely high, and the navy had to make the right decision. "On it," Kahn pointed out, "rode the very existence of the American fleet and the future of the whole Pacific war."[10]

Two Naval Intelligence officers, Commanders Joseph Rochefort and E. T. Layton, finally developed a ruse which they hoped would draw out, via Magic, the true meaning of the target cipher. They created a spurious message to be transmitted from the U.S. garrison on Midway to Naval Headquarters at Pearl Harbor. This message would "report" that Midway's freshwater distillation plant had broken down; and it would be sent by radio in the clear so that, with luck, it would be intercepted by Japanese wireless monitors. Midway complied, and Rochefort and Layton then settled down anxiously to wait. Tension mounted as the officers watched for an enemy reaction—then two days later their ingenuity and patience were rewarded: a Japanese coded intelligence message was intercepted to the effect that AF was short of fresh water because of a plant breakdown.

This, of course, eliminated all further doubt as to Yamamoto's main goal. The code detectives by that time had also succeeded in pinning down, with reasonable accuracy, the date of the enemy attack; and when the Japanese assault fleet came racing from the northwest toward Midway on June 4, 1942, Nimitz's force was properly deployed and ready. The resulting sea-air battle, grueling and complex, went on for three days and ended in an unprecedented defeat for the Japanese who lost their four finest carriers, two heavy cruisers, hundreds of planes, and over forty-eight hundred men. From that point on the forces of the United States seized the military initiative and never relinquished it.

Because of the resourceful cryptanalysts, George Marshall wrote, "We were able to concentrate our limited forces to meet their naval advance on Midway when otherwise we almost certainly would have been some 3000 miles out of place."[11]

Curiously, hoax and subterfuge were heavily involved in the two most pivotal campaigns of the war. Following Rommel's defeat at Alamein, the tide of battle in Europe turned strongly in favor of the Allies; and after Yamamoto's defeat at Midway, the military pendulum in that theater also shifted conclusively. In each of those historic watersheds, deception played a tactical role. Painted canvas in North Africa and wireless fakery in the Pacific helped in great measure to tilt the balance, and to gain Churchill's ". . . easier ways, other than sheer slaughter, of achieving the main purpose."

The prime minister's imaginative concept, however, wasn't limited in its scope solely to land fighting and naval action. There was still another dimension to the war's legerdemain, a dimension which involved the ingenious use of chemistry and electronics.

16

Coal Dust, Smoke, and Silver Foil

1940–1945

On a cool evening early in 1942, an old gentleman (his identity now lost in the yellowed file stacks of the British War Office) was walking his dog through a field on the outskirts of Coventry. Tired of tramping through the underbrush the stroller saw a smooth asphalt road up ahead, just visible in the gathering darkness. Whistling to his dog, he hurried toward the welcoming road and stepped from the grassy shoulder. There was a loud splash, a startled cry, a canine yelp of indignation. Man and animal were fished out by British soldiers camped nearby then, dried off and more or less comforted, sent on their confused way home.

The elderly gentleman and his dog had blundered into the Coventry Canal, not only astonishing themselves but adding a singular if modest footnote to what was undoubtedly the most bizarre of all British camouflage experiments in World War II.

The Coventry project involved an attempt by the Admiralty to literally camouflage water, and was motivated by then valid defense needs. Early in 1941 the Battle of Britain and the London blitz were over; but Hitler's Luftwaffe, though no longer invincible, still wielded appreciable power and throughout that year and the next, the Germans launched heavy bombing attacks on Britain's war potential, particularly her factories, supply depots, and dock areas. Colonel Turner's Q light and Starfish crews were kept very busy, and their decoys succeeded—as they had earlier—in drawing off many of the raids. But Whitehall, anxious to preserve its growing offensive strength, felt that more protection was needed. Gerald Pawle, who recorded some of these special British activities, noted:

> It became vitally necessary to protect the great factories in the Midlands. Lakes, reservoirs, and canals all helped the air navigator to pinpoint industrial centers, and it was not until the advent of radar that such visual aids lost their significance.[1]

Moonlight at that phase of the war was an airman's best ally, as its reflection on bodies of water could be seen for great distances. RAF crews on night bombing missions over Kiel could spot the Elbe River from fifty miles away; and aerial tests in England showed that critical dock areas along the Mersey could be pinpointed from thirty-five miles. "It was the heavy raids on Mersey side," wrote Pawle, "which first focused attention on the need for water camouflage, for the German bombers attacking the docks at Liverpool were observed to check their final approach by a line of five reservoirs leading from the Trent Valley."[2]

Other orientation points such as railroad lines, concrete highways, bridges, and odd patches of woodland could be—and were—often used by navigators and bombardiers; but moonlit water was the easiest, surest, and most reliable of all these visual signatures. The problem was, of course, how to eliminate this moonlight factor with effective camouflage. In Hamburg the Germans had thrown a giant wooden lid over the Binnen Alster, and while it succeeded for a time as a deception, it blocked much water traffic. Other camouflage experiments involved the use of large nets supported by cork floats, but these also limited mobility and proved practical only on ponds or small lakes.

Something else would obviously have to be found—and the Admiralty finally dumped its vexing problem into the collective lap of its Department of Miscellaneous Weapon Development, or DMWD, whose members called themselves somewhat cumbrously, "the Wheezers and Dodgers." During the war this secret unit created valuable military devices such as rocket-fired grapnels, which were used by U.S. Rangers to scale cliffs at Omaha Beach on D-day; also the great floating invasion ports known as Mulberries; and the unique RLCs, or Rocket Landing Craft—large motorized barges that could cruise parallel to a shoreline and fire huge banks of missiles at enemy installations. But undoubtedly their toughest challenge was to turn off the moon, and the Wheezers and Dodgers accepted it with unshakable aplomb.

After a great many trial runs and consultations with British chemists and fuel engineers, a DMWD camouflage team under Lieutenant Commanders Duncan Bruce, A. F. W. Coulson, and F. D. Richardson finally came up with a feasible technique: they concocted a special blend of coal dust and fuel oil which, when sprayed carefully over relatively calm water, would cling to the surface and create a dark, nonreflecting cover. The mixture, they found, could be sprayed through a high-pressure jet apparatus which worked like a vacuum cleaner in reverse.

Delighted with the potentialities of all this, Bruce and his camoufleurs decided to start right at the top and hide the noble Thames River itself.[3] The spraying apparatus was soon installed on four large launches, which chugged slowly up and down the river at night, covering the surface of the water (and the hardworking operators) with a thick sooty mantle. More than once, according to reports, a fascinated Winston Churchill strolled out on Westminster Bridge to inspect the weird results. Unfortunately a combination of wind, current and tidal flow soon tore the launch crews' filmy camouflage to shreds. Furthermore it took over one hundred pounds of the specially prepared chemical to cover a single acre of river, and the coating had to be constantly replenished. The final blow came when Thames housewives, downstream from the experiment, began to complain bitterly about the grime streaking their laundry.

This combination of quotidian tides and angry housewives proved entirely too much for the British Admiralty, and it was finally decided (despite Churchill's disappointment) to limit the eager soot-sprayers to more sheltered docksides and waterways. The most notable of these later attempts was the coating of the Coventry Canal, an orientation landmark for enemy planes heading toward war plants in the Birmingham area. Here the DMWD's camoufleurs met with success; under its dusty cloak the canal no longer glittered in the moonlight, and even at ground level it was indistinguishable from small asphalt roads in the area—a fact well known to an elderly gentleman and his surprised dog.

By that time, however, the *Luftflotten* pilots were beginning to use radar and radio-directional beams at night to guide them to their targets; bombing by moonlight was no longer a necessity, and Bruce's conjurors soon turned their attention to other and presumably less grimy activities.

Another and far more important form of chemical cover was the combat use of smoke screens. After the U.S. Army's "Torch" landings in North Africa, Ernie Pyle spent some days with a company in charge of screening a large harbor on the Algerian coast. The outfit, Pyle wrote, ". . . was part of the Chemical Warfare Service, but instead of dealing out poisonous gases they dealt out harmless smoke that covered up everything when raiders came over. By the nature of their business they worked all night and slept all day. They took their various stations in little groups in a big semicircle around the city, just before dusk, and stayed there on the alert till after daybreak." The hundred men in that particular unit, the correspondent added, "had all been together a long time and . . . had almost a family pride in what they were doing and the machinery they were doing it with."[4]

Their "machinery" consisted chiefly of fifty- and one hundred-gallon mechanical smoke generators which could, in a matter of minutes, spew out vast clouds of thick white phosphorous smoke, fog, and haze. In addition to this heavy equipment, the U.S. Chemical Warfare Service also supplied thousands of small portable smoke pots, floating pots for amphibious operations, and an endless variety of smoke grenades and munitions. Such devices provided American and British units with a simple, ever-present chemical shield; and the Office of Military History noted, "Never before had armies been able to protect their troops and hide their movements as successfully as Allied forces did in World War II."[5]

Smoke has served as a tactical weapon since the earliest of history's battles, but not until World War I did its use become standardized, particularly in the great naval engagements of that conflict. With the advent of World War II the use of smoke camouflage greatly increased, and figured in the offensive and defensive tactics not only of the Allies but of the Axis as well. The British used smoke screens early in the war, to hide many of their factories from Goering's bombers. Later the Allies used smoke to conceal their invasion beaches at Salerno, Anzio, and Elba. Artificial fog hid harbors and installations at Oran, Algiers, Bizerte, and Naples; also at Palermo, Licata and Porto Empedocle in Sicily. The Russians, after the turn of the military tide at Stalingrad, used chemical screens to hide troop and tank movements during their 1943-44 summer offensives. In the Pacific, smoke was an essential part of naval tactics. It also protected American paratroops when they made drops on New Guinea and Luzon, and helped to shield scores of U.S. island assaults.

During their early advances, the Japanese used floating generators that produced banks of smoke by the interaction of chemical mixtures and sea water. The Germans also employed smoke to hide their battleships and other targets from Allied bombers, and to cloak major troop withdrawals, such as their disengagement at Metz before Patton's Third Army. Smoke as a screen for a withdrawal action has obvious advantages, since it buys valuable time for a pullback before advancing forces can discover precisely what action the enemy is taking.

As the war continued, small efficient smoke generators were mounted by the belligerents on aircraft and motor-torpedo boats, on tanks, armored gun carriers, even on jeeps and scout cars. One late-model German panzer tank, the KW-3, had no less than six smoke dischargers bolted to its menacing turret; these tilted outward to create a fanlike gray cloud in front of the vehicle, for use when under attack.[6]

Smoke screening is usually considered a passive or defensive measure, but its greatest and most dramatic use in World War II came in connection with an offensive operation.

In February 1945, Allied armies were poised along the west bank of the Rhine River—the last great barrier to the complete overrunning of the Third Reich. On the east bank the Wehrmacht divisions, weary and demoralized but still determined to resist, gathered their strength for a final all-out, desperate defense. Even with their great superiority of men, arms, and equipment, crossing this vast river would be difficult for the Allies. "The Rhine," Eisenhower wrote, "was a formidable military obstacle, particularly so in its northern stretches. It was not only wide but treacherous, and even the level of the river and the speed of its currents were subject to variation because the enemy could open dams along the great river's eastern tributaries."[7]

By a fortuitous, unexpected stroke of Allied luck, the Germans in their hasty withdrawal had left the Ludendorff Bridge standing at Remagen; and units of General Courtney Hodges's First Army pushed across it quickly to establish a solid bridgehead. But this lone outpost had to be supported; and in order to insure success, Eisenhower planned to breach the river barrier along its entire length. Hodges was already across at Remagen. South of that point the river narrowed and effective crossings could be handled by the U.S. Third and Seventh Armies, and the First Army of the Free French. In the north, midway between Duisberg and Arnhem, an assault would be made by Montgomery's British forces and the U.S. Ninth Army under General William Simpson. Here, where the river was widest and the defenders waited in strength, a crossing would be the most difficult.

To increase the Ninth Army's chances of surprise, a fake "amphibious buildup" was created in the Dusseldorf area, well south of the genuine troop concentrations.[8] Engineer camouflage units built dummy barges, fuel dumps, bridge-building depots, and tank assembly parks—using for this display many of the inflatable rubber tanks that had given such helpful service in the Skye and Quicksilver deceptions before D-day. The diligent camoufleurs even created a large dummy medical facility to take care of the "battle casualties" which would inevitably occur. Real Medical Corps ambulances were rerouted through this area to give it added credibility.

Farther north, the genuine preparations for the crossing were carried out under strict camouflage discipline. To better conceal this huge buildup and hide it from German artillery, Montgomery and Simpson decided to use smoke cover. But a smoke screen concentrated at one focal area of attack would mean giving away the Allies' plans to the enemy, or would at least rouse German suspicions and make them all the more watchful. The solution to the problem was as simple as it was unprecedented: The commanders decided to use smoke coverage *over the entire front* of the British Second Army and the Ninth Army's XVI Corps opposite Wesel, preventing the Germans from guessing where, if anyplace, the actual northern blow would fall. As a result—whenever visibility conditions warranted—a giant curtain of fog was lowered over the entire west bank of the river, stretching for more than *twenty miles*, and cloaking a continuous strip of some 6,800,000 square yards.

To achieve this huge cover—never before attempted on such a scale—the Allies set up a special Smoke Control Center, used the manpower of a whole Chemical battalion, rounded up every smoke pot and generator for many miles around, and commandeered dozens of extra trucks to transport the equipment to the right places. Fast motor launches carrying smoke-making apparatus were also assigned to roar along the river's bank during the operation.

The prodigious effort proved its value. On the night of March 23, British and American units, after a heavy opening bombardment by planes and artillery, launched their drive across the lower Rhine. Airborne troops, using 1,570 planes and 1,320 gliders, participated in the assault—an action characterized by Eisenhower as "the most successful airborne operation we carried out during the war."[9] The Germans, stunned by the initial barrage, thrown off balance by the dummy concentrations, and confused by the massive smoke screen, failed to properly mobilize their forces; and by noon of the following day the Allies had won and consolidated a bridgehead on the far shore of Hitler's last and greatest defensive barrier.

In addition to chemical subterfuge, World War II was marked by an amazing variety of electronic ruses, all of which—though not in the immediate purview of military camouflage—played their part in a consistent policy of "disguise and divert."

The war years saw the development of radar, sonar, asdic, and many other high-frequency directional aids (known colloquially as "huff duff"); but as quickly as these were perfected, countermeasures were created to oppose them. During the London blitz, the Luftwaffe had begun using a new system of night bombing based on electronic guidepaths. Their device—code-named "Knickebein" by the Germans—consisted of a double radio beam that acted like an infallible set of aerial railroad tracks. By flying right down the center of this split beam, which was transmitted from towers behind the Channel coast, a bomber pilot would be led accurately to his goal —*regardless* of visibility conditions. Later, two such split-beam transmitters could be set up at widespread points so as to intersect over the target. The Luftwaffe flier, following along one beam, had merely to watch for this intersection signal on his instrument panel, then drop his bombs with almost certain success.

"No longer, therefore," Churchill wrote of the Knickebein apparatus, "had we only to fear the moonlight nights, in which at any rate our fighters could see as well as the enemy, but we must even expect the heaviest attacks to be delivered in cloud and fog."[10]

This soon touched off a covert electronic struggle which the prime minister sometimes referred to as "the Battle of the Beams." Technicians of the British Air Ministry, with support from scientists such as Drs. R. V. Jones and Frederick Lindemann, went diligently to work to find a method of neutralizing the Nazi threat, and before long they created a technique for jamming the enemy's beam transmissions. But Churchill wasn't satisfied. Jamming, if successful, would be immediately evident to the Luftwaffe pilots, and the Germans would quickly develop countermeasures of their own.

Searching for a more subtle approach, the British experts finally succeeded in *warping* the German radio beams so that unsuspecting fliers would be deflected some ten or twenty miles from their true path; as a result Goering's pilots, flying on cloudy overcast nights, often bombed empty fields under the impression that they were right over their targets, and tons of enemy explosives were wasted in this manner.

Through 1941 the Germans made constant improvements in their radio-directional systems, but the British managed to counteract all these and to continue deflecting the beams. Typical of their success was a German night raid in May of that year on the Rolls-Royce aircraft engine plant at Derby. The Luftwaffe attack was a heavy one, and Churchill reported afterward with obvious relish:

> The German communiqué claimed the destruction of the Rolls-Royce Works at Derby, which they never got near. Two hundred and thirty high-explosive bombs and a large number of incendiaries were, however, unloaded in the open country. The total casualties there were two chickens.[11]

Other ruses and deceptions involved radar, the most significant of the war's electronic developments. Used initially by the British, radar—an acronym for "radio detecting and ranging"—was quickly adopted by all the belligerents, and proved invaluable in giving advance warning of the approach of enemy aircraft and naval units. It could also supply data as to the size, range, and bearing of the approaching forces.

Improvements in radar equipment and techniques came very rapidly and so, too, did the inevitable countermeasures. Chief among these was a simple jamming device called "Window," also known as "Chaff," which consisted of thin strips of coated metallic foil, resembling the silver foil used as Christmas tinsel decoration. These silvery strips, when floated through the air in large quantities, appeared on the cathode-ray tube of a radarscope as a blizzard, obscuring all normal signals. During the war, and particularly prior to Neptune, bales of Window were packed aboard long-range bombers and literally shoveled out into the planes' slipstreams, to confuse and mislead German intelligence units.

This unique concept was the brainchild of Britain's TRE (Telecommunications Research Establishment) and a secret U.S. research unit based at Harvard—"Laboratory 15"—whose special committee was headed jointly by Drs. Robert Cockburn and Joan Curran. Working under great pressure, the committee studied and analyzed German radar equipment that had been captured during an early commando raid, and soon came up with Window and other jamming techniques for disrupting the pulse and frequency modulations of enemy radar screens.

Later Dr. Curran, a prestigious scientist, discovered an even more valuable use for Window: She found that by dropping the metallic foil in carefully controlled, prearranged patterns, it would create the exact impression on a radarscope of a genuine air squadron. She also developed a mobile electronic device, code-designated "Moonshine," which could be installed directly aboard a ship or plane. The device reflected radar pulses in such a way as to simulate the approach of large fleets of warships and aircraft. In April 1942, Moonshine was first used to lure the Luftwaffe away from a planned Eighth Air Force attack, and the test was successful. Several RAF Defiants equipped with the instrument, Cave Brown reports, ". . . circled over the Thames to duplicate the pulses created by bomber streams assembling for a raid." According to signals intercepted by Ultra, ". . . the Luftwaffe controller launched one hundred and forty-four fighters to meet the deception thrust."[12] As a result, less than half that number of German planes were available to oppose the genuine Allied raid which was aimed at Rouen.

Window, Moonshine, and similar inventions lent themselves perfectly to ruse and misdirection, and were used with great success to confuse the Nazis and shield Allied combat operations, especially during the Normandy landings and the accompanying feints by phantom FUSAG divisions supposedly aimed at Calais.

Among other electronic hoaxes—employed at times by the Axis, but far more effectively by the Allies—were fake radio traffic, wireless jamming, decoy messages to nonexistent units, cryptographic feints, spurious signals to the underground, and the counterfeiting of transmissions from secret agents. The forging of these transmissions, an important intelligence strategem, was based on the remarkable ability of X-2 and MI-5 experts to duplicate the radio style, or "fist," of captured spies, so that their messages could be convincingly imitated when necessary.

All this, which Brown called electronic spoofery and

which Churchill referred to as the wizard war, has been documented in numerous accounts dealing with espionage, communications, and cryptography. It suffices to simply note here that electronics added a new nonvisual dimension to the ancient art of military subterfuge. Electromagnetic weaponry—created by science and unknown in previous conflicts—was made part of the enormous Allied arsenal, and became an appreciable factor in the successful planning of the campaigns.

17

Phantasm and Reality: An Epilogue

The illusion was startling, audiences were stunned and mystified, and Harry Houdini became known not only as a great escape artist, but as "the man who could walk through walls."[1]

As the presentation begins the spectators see an empty stage covered by a large rug. Over this rug a seamless canvas cloth is spread. At the very center of the stage, on a flat steel beam, a team of bricklayers rapidly builds a thick, solid brick wall. Observers from the audience have been invited by Houdini to come up on stage and watch this work closely. The brick wall, with one end facing the audience, is one foot thick, ten feet long, and eight feet high. It has been built in full view of everyone, and is unquestionably solid. Next, two narrow three-wing folding screens are carried out and placed against the center of this wall, one on each side. The committee members take up positions around the entire display, their feet planted firmly on the canvas stretched over the carpet.

Now Houdini, with a courtly bow, steps behind one of the screens and waves his hand over the top. There is a fanfare, a ruffle of drums from the orchestra, and moments later Houdini's hand can be seen waving above the second screen, on the other side of the brick barrier.

The magician steps out to acknowledge the applause of his audience, which is standing and cheering enthusiastically—many of them naively convinced that the performer has actually "dematerialized" himself in order to reappear on the wall's far side.

The method Houdini used to "walk through walls" was simplicity itself, and decidedly anticlimactic. In the wooden floor of the stage, directly under the steel beam and heavy wall, was a small trapdoor. As soon as Houdini stepped behind the first screen, this trapdoor was opened. There was just enough slight sag in cloth and carpet to allow the agile magician to wriggle underneath quickly and come up on the other side. The committee members, with their feet planted on the cloth, kept the floor coverings from sagging noticeably, and the closing of the trapdoor eliminated any telltale bulges. Afterward, the observers were invited to examine every inch of brick wall and steel beam but, of course, nothing suspicious could be discovered.

The dramatic success of that particular illusion was based not on its simple sub rosa mechanics, but on the visual impact it achieved. Similarly, the illusions created by the magi of World War II depended for success on the *effects* they produced, which in turn influenced the tactics and intelligence evaluations of the enemy.

Military objectives throughout the war led to the need for disguising tactical plans and misdirecting the enemy's attention; and meeting this need called for aural deception, electronic legerdemain, and shrewd visual sophistry. In every case the *purpose* was the same, whether it involved Jo Mielziner's artists spray-painting an airfield runway, John Turner stringing acres of Q lights, Barkas hiding Montgomery's tanks under their Sunshields, Lockheed's camoufleurs simulating a quiet California suburb, FUSAG wireless operators creating fake radio traffic, Everett Warner designing ship dazzle patterns, MacArthur's gun crews draping their howitzers with nets and branches, Maskelyne launching his canvas submarines, or Ewen Montagu conjuring a *vita verus* for an anonymous corpse. These and scores of other conjuries worked toward one common goal: to fool the enemy and create military odds that would be more favorable to the Allies.

Camouflage methods developed in World War II were carried over in later conflicts. During the Korean War of 1950–53, the UN forces put heavy emphasis on field decoys and dummies such as fake tanks, aircraft, and artillery. In the Arab-Israeli confrontations, both sides made wide use of disruptive and sand-colored paint for their tanks and half-tracks. And in Vietnam, with its grim overtones of earlier jungle campaigns against the Japanese, particular attention was paid to camouflage clothing, individual concealment, and the use of natural cover.

In the gruesome and unthinkable context of a future war between major powers, known forms of field camouflage would be meaningless. Armageddon is irreducible, and there are no simple ways to fool ballistic missiles carrying multiple nuclear warheads. Modern aerial and satellite reconnaissance methods are also far less open to deception. Infrared photographs, with their sensitivity to plant chlorophyl, can instantly detect artificial foliage. Army movements under cover of darkness (so vital in World War II) have been rendered impossible by night photography techniques that light up a scene as vividly as the brightest sunshine. And high-altitude supersonic "drones" equipped with rotating optical units now provide horizon to horizon photo cover, producing accurate prints that can delineate a golf ball on a putting green from ten miles in the air.

Most important of all, polar orbiting recon programs such as the U.S. Air Force's SAMOS (Satellite and Missile Observation System), and equivalent satellite devices being used by the U.S.S.R., can photograph whole continents with incredible clarity and detail.[2] In addition, new space systems are being developed that operate in the electromagnetic spectrum. These sensors can "see" radiation and spectral signals in terms of wavelengths completely invisible to the human eye. In the foreseeable future, the holographic process, involving laser beams, will also be used, and will be able to photograph targets and suspicious areas in full three dimensions, at any required scale.

There are undoubtedly, in today's arsenals, numerous electronic devices for misdirecting and negating rockets and missiles, and for neutralizing satellite reconnaissance cameras; but these lie in complex areas far beyond the human-scale deceptions and simple technologies that so often proved of value during the war against the Axis.

In the contemporary flood of World War II military histories, analyses, battlefield chronicles, personal memoirs and diaries, the function of the camouflage units has been, with some few exceptions, largely overlooked; yet both sides during the fighting used camouflage techniques to great advantage. It is hoped that this record will help somewhat to fill the gap and place the work of the camoufleurs in its proper context. Of course any account which deals with a single facet of military history runs a built-in risk of advocacy, since there is always the tendency to exaggerate the importance of a particular element and thus distort its contributions. The war was certainly not won by camouflage, espionage, cryptanalysis, aerial bombardment, logistical supply, command strategy, or even by the Allies's great industrial output; but by the soldiers, sailors, airmen, and marines who faced danger, fought, persevered, endured, suffered, bled, and died. The triumphs and the accolades belong to them; all other factors, however valuable, played essentially supportive roles.

Camouflage, in this supportive context, added its own unique leverage; the Houdini touch permeated the war's battles, and "fool the eye" was applied more frequently than in any conflict in history. When Homer St. Gaudens went overseas as a young camouflage officer in World War I, many American doughboys naively asked if he could supply them with magic "invisibility paint." By World War II all this, of course, had changed radically, and levels of sophistication matched the brilliance of the cover techniques. In addition, the hoaxes practiced by the Allies—particularly in England and North Africa—were so effective that they caused great confusion among German Intelligence units and consequent distrust on the part of Hitler. Albert Speer has noted that in the weeks before D-day, the Nazi dictator received a flood of ". . . contradictory predictions on the time and place of the invasion from the rival [German] intelligence organizations."[3] These contradictions, planted and nurtured by the careful fictions of Fortitude, led to the führer's disenchantment with the *Abwehr,* von Roenne's FHW, and all his other intelligence sources.

By the time Neptune was under way, Speer added, Hitler ". . . scoffed at the various services, calling them all incompetent and, growing more and more heated, attacked intelligence in general."[4] All of which (fortunately) limited his capacity to make cogent and meaningful decisions.

"Military history and drama," wrote Hanson Baldwin, "are compatible; in fact, they are inseparable; one without the other is incomplete."[5]

In the vast drama that unfolded between 1939 and 1945, the camouflage troops were among the many bit players. But with stage props of canvas, nets and chicken wire, lumber, smoke, paint pots, and palm fronds, they performed their special kind of battlefield wizardry, helping in some measure to protect lives and to hasten, however slightly, the final Allied victory.

Glossary

AA: Antiaircraft artillery.

Abwehr: German intelligence and counterespionage service under Admiral Canaris.

AF: Japanese code designation for Midway Island, target of Yamamoto's carrier force attack.

A-Force: Deception arm of the LCS in the Middle East and Mediterranean area.

Aldis lamp: Hand-held signal lamp used for flashing messages to ships and aircraft.

Amt Mil: German intelligence-gathering service under General Schellenburg.

Armeegruppe Patton: Name given by German intelligence to Allied force they believed was preparing to invade the Pas de Calais.

Asdic: Early antisubmarine detection apparatus—a forerunner of sonar.

ATS: Auxiliary Territorial Service (British Army).

Bertram: Cover and deception plan used by the British prior to the battle of El Alamein.

Bodyguard: Code name for the overall Allied deception plans covering the Normandy landings.

C in C: Commander in Chief.

COE: Corps of Engineers (U.S. Army).

Copperhead: North African deception scheme involving the use of a double for Marshal Montgomery.

Crossbow: Code name for Allied air campaign against enemy V-bomb and missile sites.

Crusader: British military operation in North Africa which lifted the siege of Tobruk.

CSM: Committee of Special Means (SHAEF cover and deception unit, linked with the LCS).

CWS: Chemical Warfare Service (U.S. Army).

Dazzle painting: Disruptive colors and patterns used on ships during war to create course distortion.

Decoy: Fake object or installation deliberately displayed as a target to draw enemy fire.

Diamond: Code name for dummy water pipeline built in the desert as part of Operation Bertram.

Dicing: Descriptive term for highly dangerous, low-level air reconnaissance missions (literally, "throwing dice").

DMWD: Department of Miscellaneous Weapon Development (British).

Drone: Small pilotless aircraft operated by remote-control radio, used in long-range reconnaissance.

Dummy: Fake object or installation displayed partly as a target, but chiefly to mislead the enemy and create a false impression.

Feldgrau: A term for the average German soldier (literally, "field-gray").

Fist: Colloquial term for the distinctive individual touch and message style of a wireless radio operator; i.e., the operator's handwriting.

Flat-top: A camouflage net stretched flat and horizontal over an object to be concealed.

Fliegerhauptmann: German Flight Captain (squadron leader).

Fortitude: Code name for overall Allied deceptions designed to mislead German forces in Norway and France.

Fremde Heere West (FHW): German intelligence-gathering service on the Western front, under Colonel von Roenne.

Freya: Type of field radar unit used extensively by the Wehrmacht.

FUSAG: First U.S. Army Group, a largely fictitious organization used in pre–D-day Fortitude deceptions.

G-1: Personnel and administration (U.S. Army General Staff).

G-2: Intelligence activities (U.S. Army General Staff).

G-3: Operations and training (U.S. Army General Staff).

G-4: Supply and maintenance (U.S. Army General Staff).

GCI: Ground-controlled interception, used in directing aircraft toward targets.

Huff duff: Colloquial term for high-frequency radio direction-finding techniques.

Husky: Code name for the Allied invasion of Sicily.

Joint Chiefs of Staff: The U.S. military high command, Washington, D.C., headed by General Marshall.

Junker: Member of elite Prussian aristocratic and military caste, powerful in pre-WW II Germany.

K sites: Decoy airfields created by the British to draw off Luftwaffe raiders.

Knickebein: German code name for radio-directional beams used by Luftwaffe bombers.

Kriegsmarine: German Navy, under Admiral Raeder and subsequently Admiral Doenitz.

LCS: London Controlling Section, responsible for directing all Allied cover and deception activities.

LCT: Landing Craft Tank—a barge designed to hold four Shermans.

Lighter: Harbor vessel chiefly used for carrying and unloading cargoes from ship to shore.

Lightfoot: Code name for British military offensive launched at El Alamein.

Luftflotten: German air fleets, or squadrons.

Luftwaffe: German Air Force, under Goering.

Meillerwagen: Mobile units used by the Germans to service and transport their V-2 rockets.

MI-5: British counterespionage and security service.

MI-6: British intelligence service responsible for overall covert operations (counterpart of the OSS).

Mincemeat: Code name for Allied deception plan to disguise the assault on Sicily.

Moonshine: Apparatus designed by the Allies to simulate ships and aircraft on enemy radarscopes.

Neptune: Allied code name for the assault on the Normandy beaches (part of Overlord).

Oberkommando der Wehrmacht (OKW): The German military high command.

OCE: Office of the Chief of Engineers (U.S. Army).

OMH: Office of Military History, Washington, D.C.

Osnaburg: Dyed cotton canvas strips used for weaving into camouflage nets.

OSS: Office of Strategic Services (U.S.).

Overlord: Allied code name for the military operational plan to invade and liberate western Europe.

PI: Photo Interpreter (U.S. and British Air Forces).

Q lights: Decoy lighting patterns used in various Allied deceptions.

Quicksilver: Allied deception plan prior to D-day, also known as Fortitude South.

RAF: Royal Air Force (British).

Recce: Colloquial term for Allied reconnaissance activities, both photographic and visual.

SAMOS: Satellite and Missile Observation System (U.S. Air Force).

Scrim net: Light, gauzy netting with very fine mesh which can be painted; sometimes used in camouflage deceptions.

Sea Lion: German code name for their (unrealized) plan to invade and conquer England.

SHAEF: Supreme Headquarters Allied Expeditionary Force, headed by General Eisenhower.

Sicherheitsdienst (SD): Intelligence and counterespionage branch of the Nazi SS.

Skye: Allied deception plan prior to D-day, also known as Fortitude North.

SS: Schutzstaffel, the military and political arm of the Nazi party.

Staffel: Luftwaffe air squadron.

Starfish: Decoy fire baskets used for deception by English defenders during the Battle of Britain.

Stereo viewer: Optical device enabling user to produce three-dimensional effects with overlapping photographs.

Sunshields: Camouflage truck covers, used for hiding British tanks prior to Operation Lightfoot.

Torch: Code name for Allied military landings in North Africa.

UHF: Ultrahigh frequency radio and electronic communications.

Vergeltungswaffen: German vengeance or "reprisal" weapons designed to terrorize Britain, and eventually the U.S.

Wehrmacht: The military forces of Germany (literally "war power").

Window: Code name for silver foil dipoles used to jam enemy radar.

X-2: Branch of OSS specifically responsible for enemy agent control and counterespionage.

XX Committee: Arm of MI-5 responsible for developing and controlling double agents.

Notes and Sources

1
IMAGE AND ILLUSION: A PROLOGUE

1. . . . is based on deception." This and subsequent quotation from *The Art of War* by Sun Tzu, cited in *Encyclopedia Britannica*, 1969, Vol. 4, pp. 708–711.

2. Hoax and deception . . . *et seq*. Earlier uses of military misdirection are noted in Sir Ronald Wingate's *Not in the Limelight* (London: Hutchinson, 1959), pp. 190–191.

3. In the natural world . . . *et seq*. For a scientific, comprehensive account of camouflage in the animal kingdom, see Adolf Portman's *Animal Camouflage* (Ann Arbor: University of Michigan Press, 1959).

4. . . . by any warring nation previously." *Britannica, op. cit.*, Vol. 4, p. 709.

5. . . . from the camouflage itself." Constance Babington-Smith, *Air Spy* (New York: Harper and Bros., 1957), p. 113.

6. . . . of achieving the main purpose." Winston Churchill, *The Great War* (London: George Newnes, 1933), Vol. 1, p. 498.

2
BODYGUARD—THE HOUDINI TOUCH

1. . . . puzzled as well as beaten." Churchill, *op. cit.*, p. 498.

2. Within a few months . . . *et seq*. For a detailed account of Rom-

mel's efforts to reinforce the Atlantic Wall, see Desmond Young's *Rommel the Desert Fox* (New York: Harper, 1950), Berkley Medallion edition, pp. 189–199.

3. . . . emerging as master of Europe." Anthony Cave Brown, *Bodyguard of Lies* (New York: Harper & Row, 1975), p. 3. Brown's prodigious history has proven a particularly helpful source of information, to which I am much indebted, on the development and implementation of the LCS and Fortitude strategies prior to D-day.

4. The German dictator . . . *et seq. Ibid.*, p. 426.

5. In 1943, German factories . . . *et seq. Ibid.*, pp. 424–425.

6. . . . is bound to fail." Gilles Perrault, *The Secret of D-Day* (Boston: Little, Brown, 1965), p. 6.

7. . . . the whole works on one number." Brown, *op. cit.*, p. 614.

8. . . . a fifty-fifty chance." *Ibid.*, p. 614.

9. . . . while it is still afloat." Young, *op. cit.*, p. 200.

10. . . . and place of the invasion." Brown, *op. cit.*, p. 432.

11. . . . they think they are watching *you*." Walter B. Gibson, *The Original Houdini Scrapbook* (New York: Sterling, 1976), p. 173, article written by Houdini in 1919.

12. . . . altogether outside his calculations." Lieutenant General Sir Frederick Morgan, *Overture to Overlord* (New York: Doubleday, 1950), p. 239.

13. The Fortitude ruse . . . *et seq.* SHAEF Ops. "B" document #18209, quoted by Brown, *op. cit.*, p. 460.

3
THE PHANTOMS OF FORTITUDE

1. . . . in its final result." Brown, *op. cit.*, p. 426.

2. It could, the Fuehrer pointed out . . . *et seq. Ibid.*, p. 426.

3. . . . bound for Norway. *Ibid.*, p. 465.

4. . . . as to your intentions." John C. Masterman, *The Double-Cross System* (New Haven: Yale University Press, 1972), p. 7. Masterman documents the unique work of Britain's XX Committee from 1939 to 1945, as well as specific preinvasion deceptions.

5. . . . from a particular area." *The Secret War Report of the OSS*, ed. by Anthony C. Brown (New York: Berkley Medallion Books, 1976), p. 95.

6. . . . espionage system in this country." Masterman, *op cit.*, p. 3.

7. . . . also messages sent to him." *Ibid.*, Foreword by Norman H. Pearson, page x.

8. . . . for a successful stroke." *Ibid.*, p. 22.

9. In the United States . . . *et seq.* J. Edgar Hoover's recounting of the "Albert van Loop" episode appears in *Secrets and Spies* (New York: Reader's Digest, 1964), pp. 283–287. However, Ladislas Farago, in *The Game of the Foxes* (New York: Bantam Books, 1971), pp. 828–838, presents his own evaluation of "van Loop" as in fact a triple agent.

10. . . . where *I* would attack." Perrault, *op. cit.*, p. 160.

11. . . . weapons were concentrated." *Ibid.*, p. 161.

12. . . . expected the Allies." *Ibid.*, p. 165.

13. . . . by these circumstances." Masterman, *op. cit.*, p. 146.

14. . . . 'Old Blood and Guts'." Ladislas Farago, *Patton: Ordeal and Triumph* (New York: Dell, 1963), p. 12.

15. At that point . . . *et seq.* The strange, unequal battle between bull and rubber tank has been noted by Brown, *op. cit.*, p. 603.

16. . . . disclose the whole plan." Public Record Office, London. SHAEF file # 37/882 (AIR).

17. . . . take them to Calais?" Brown, *op. cit.*, p. 604.

18. . . . especially Garbo and Brutus." Masterman, *op. cit.*, p. 156.

19. . . . where the Allies would land." Brown, *op. cit.*, p. 220.

20. . . . landing in that area." Perrault, *op. cit.*, p. 220.

21. . . . would be rendered difficult." Dwight D. Eisenhower, *Crusade in Europe* (New York: Doubleday, 1948), p. 249.

22. . . . first week of June." Brown, *op cit.*, p. 639.

23. . . . was upon them." Walter Warlimont, *Inside Hitler's Headquarters* (New York: Praeger, 1966), p. 422.

24. . . . that has ever taken place." Winston S. Churchill, *The Second World War* (Boston: Houghton Mifflin, 1953), Vol. 6, p. 6.

25. . . . cover and deception techniques." Hanson W. Baldwin, *Battles Lost and Won* (New York: Harper & Row, 1966), Avon edition, p. 356.

26. . . . sheer weight of numbers." Perrault, *op. cit.*, p. 239.

27. . . . different place anyway." Brown, *op. cit.*, p. 659.

28. . . . this confused thinking." Churchill, *Second World War*, Vol. 6, p. 11.

29. The Nazi dictator . . . *et seq.* Hitler's evaluation of Neptune

recalled by Albert Speer, *Inside the Third Reich* (New York: Macmillan, 1970), Avon edition, pp. 455–456.

30. . . . single hoax of the war." Omar N. Bradley, *A Soldier's Story* (New York: Henry Holt, 1951), p. 344.

31. . . . all of it completely real." Chester Wilmot's figures quoted by Baldwin, *op. cit.*, p. 355.

4
Q LIGHTS, FIREWOOD, AND OLD RUBBER TIRES

1. The letter to RAF . . . *et seq.* Dated 22 July 1940, British Public Record Office file #2/3212. Many facts re Colonel Turner's activities were uncovered in documents made available to me at the British Public Record Office (Air Ministry) in London, off Chancery Lane. These PRO files, declassified only in the past few years, tell a remarkable story which, until this writing, has been largely unknown to the American and British publics.

2. . . . not won by evacuation." Churchill, *Second World War*, Vol. 2, p. 115.

3. In France . . . *et seq. Ibid.*, Vol. 2, p. 141.

4. . . . across the narrow Channel." William L. Shirer, *The Rise and Fall of the Third Reich* (New York: Simon and Schuster, 1960), p. 737.

5. Turner's first decoys . . . *et seq.* PRO #2/6021.

6. As had been feared . . . *et seq.* PRO #39/112.

7. . . . his lack of success." From letter dated 9 July 1940, PRO #2/3212.

8. Turner's Starfish . . . *et seq.*, PRO #2/4759.

9. . . . against the aircraft industry." From *Hitler's War Directives, 1939–1945*, ed. by H. R. Trevor-Roper (London: Pan Books, 1964), p. 79.

10. . . . to their efficiency." Basil Collier, *The Battle of Britain* (New York: Berkley Medallion, 1969), p. 96.

11. To help build the K sites . . . *et seq.* PRO #2/3212.

12. . . . my best decoys!" Interview with Robin Brown.

13. on the Muritz Zee . . . *et seq.* PRO #34/629.

14. The day finally came . . . et seq. Several versions of this anecdote exist, the most reliable of which can be found in Major M. E.

DeLonge's *Modern Airfield Planning and Concealment* (New York: Pitman, 1943), p. 135.

15. . . . of large British cities." Collier, *op. cit.*, p. 105.

16. . . . did not succeed." Churchill, *Second World War*, Vol. 2, p. 342.

17. . . . wonderful organization." *Ibid.*, p. 342.

18. The Q and K . . . *et seq. Britannica, op. cit.*, Vol. 4, p. 709.

19. . . . achieved remarkable results." Churchill, *Second World War*, Vol. 2, pp. 388–389.

20. . . . in the English-speaking world." *Ibid.*, p. 359.

5
"NO FUNDS AVAILABLE"

1. . . . no need for concealment." Barbara W. Tuchman, *The Guns of August* (New York: Macmillan, 1962), p. 37.

2. . . . of all colors." *Ibid.*, p. 38.

3. . . . you can imagine." Brown, *op. cit.*, p. 72.

4. . . . army of World War I." Eisenhower, *op. cit.*, p. 7.

5. Modern air attack . . ." *et seq.* This and subsequent citations re General Robins's efforts to obtain camouflage protection for U.S. airfields are from *U.S. Army in World War II—The Corps of Engineers: Construction in the U.S.* by Lenore Fine and Jesse A. Remington (Washington, D.C.: Office of the Chief of Military History, 1972), pp. 448–450.

6. One installation in the Midlands . . . *et seq.* "British Civil Camouflage Practice." Report to OCE by Lieutenant Colonel Homer St. Gaudens, 12 Feb. 1942, pp. 8–11.

7. . . . has yet been bombed." *Ibid.*, p. 13.

8. . . . cost-estimated at $56,210. Letter from Lieutenant General W. S. Short to the Adjutant General, Washington, D.C., 12 July 1941. U.S. National Archives, RG #18, Series II (Engr. 000.91).

9. In spite of the . . . *et seq.* Memos and notes re early camouflage efforts from National Archives, *ibid.*, RG #18.

10. . . . in this department." Short, *op. cit.*, NA.

11. . . . to a collison course." Hanson W. Baldwin, *The Crucial Years, 1939–1941* (New York: Harper and Row, 1976), p. 354.

12. . . . against the U.S.A." Shirer, *op. cit.*, p. 879.

13. . . . and, as he said, 'severely.' " *Ibid.*, p. 871.

6
OUT OF THE COCOON

1. They dove out . . . *et seq.* Many detailed accounts are available on the Japanese surprise attack on Pearl Harbor. Most of the facts given in this brief chapter are from Baldwin's *The Crucial Years*, pp. 365–372.

2. . . . wingtip to wingtip." *Ibid.*, p. 363.

3. . . . hard to comprehend." Captain Ellis M. Zacharias, *Secret Missions* (New York: G.P. Putnam, 1946), p. 251.

4. . . . of all time." Baldwin, *The Crucial Years*, p. 370.

5. . . . formal state of war." *New York Times*, 8 December 1941, p. 1:1.

6. . . . doubt about the end." Churchill, *Second World War*, Vol. 3, p. 607.

7. . . . wider dispersal of planes." Fine and Remington (OMH), *op. cit.*, p. 478. General Arnold's directive dated 9 Dec. 1941.

8. . . . our national capacity." *Ibid.*, p. 480.

7
DISGUISE AND DIVERT

1. "Once the fire . . . *et seq.* Churchill, *Second World War*, Vol. 3, pp. 607–608.

2. . . . implements of war." Fine and Remington (OMH), *op. cit.*, p. 480.

3. . . . a blank check." *Ibid.*, p. 478.

4. At that point . . . " *et seq.* Interview with William Pahlmann.

5. . . . grasping it at all." *New York Times*, 16 March 1976, p. 38:1.

6. . . . obtains his results." Also directly preceding. Brown, *op. cit.*, p. 46.

7. . . . of the Pacific States." Stetson Conn, Rose C. Engelman, and Byron Fairchild, "The Western Hemisphere—Guarding the U.S. and Its Outposts" (Office of the Chief of Military History, 1964), p. 82.

8. During February . . . *et seq. Ibid.*, p. 87.

9. U.S. camouflage projects . . . *et seq.* OMH, *History of Western Defense Command*, Vol. 3, chap. XIV.

10. six tons per plane . . . *et seq.* Harold E. Wessman and William

A. Rose, *Aerial Bombardment Protection* (London: Wiley and Sons, 1942), p. 3.

11. . . . entire industrial areas." Article, "Camouflage in Wartime," from *Lockheed-Vega Aircraftsman,* June 1943, p. 69.

12. The Lockheed-Vega . . . *et seq. Of Men and Stars* (Lockheed Aircraft Corp., Aug./Sept. 1957), pp. 3–4.

13. . . . their earlier conquests." Conn, Engelman, and Fairchild (OMH), *op. cit.,* p. 92.

14. . . . attract a pilot's eye." Ernie Pyle, *Here Is Your War* (New York: Holt, Rinehart and Winston, 1971), p. 142.

8
EVERYTHING BUT THE AMMUNITION

1. . . . with bow and arrow." Stefan Lorant, *The New World* (New York: Duell, Sloan and Pearce, 1946), p. 85.

2. . . . or conform to it." Geoffrey Barkas, *The Camouflage Story* (London: Cassell and Co., 1952), p. 43.

3. . . . but the ammunition." Interview with Irwin Greenberg.

4. As an odd sidelight . . . *et seq.* Interview with William Manson.

5. . . . win the next war." Andrew J. Brookes, *Photo Reconnaissance* (London: Ian Allen, 1975), p. 35.

9
"LET THE PHOTOS SPEAK . . ."

1. . . . each plate after that." Brookes, *op. cit.,* p. 19.

2. . . . some 13 million prints. U.S. Air Force official field report, "Reconnaissance in the Ninth Air Force," 9 May 1945, p. 49.

3. . . . engine was lost." Glenn B. Infield, *Unarmed and Unafraid* (New York: Macmillan, 1970), p. 87.

4. . . . into the junk pile." *Ibid.,* pp. 96–97.

5. . . . notable books. Among Saint-Exupéry's successful volumes are *Night Flight* (New York: Century, 1932), *Wind, Sand and Stars* (New York: Reynal and Hitchcock, 1939), and *Flight to Arras* (New York: Reynal and Hitchcock, 1942). He is also the author of the noted children's fantasy, *The Little Prince* (New York: Harcourt, Brace and World, 1943).

6. Over 1,700 T Reports . . . *et seq.* USAF field report, *op. cit.*, p. 63.

7. In the summer of . . . *et seq.* Babington-Smith, *op. cit.*, pp. 105–107.

8. So to divert . . . *et seq.* Diagrams and a description of this elaborate construction can be found in K. F. Wittman's *Industrial Camouflage Manual*, prepared for Pratt Institute, Brooklyn, N.Y. (New York: Reinhold, 1942), pp. 119–121.

9. . . . of Allied lives." Eisenhower, *op. cit.*, p. 323.

10. . . . 720 separate locations. Brookes, *op. cit.*, p. 188.

11. . . . blades of grass." Babington-Smith, *op. cit.*, p. 191.

12. Recon planes of . . . *et seq.* USAF field report, *op. cit.*, p. 55.

13. . . . a matter of hours." Eisenhower, *op. cit.*, p. 453.

10
THE DEFENSE OF TARGET 42

1. . . . almost totally demolished. *Cf.* Brown, *op. cit.*, pp. 720–721.

2. . . . incredible speed and power. Perhaps the best and most clearly detailed of the many accounts of the *Vergeltungswaffen* is Basil Collier's *The Battle of the V-Weapons* (London: Hodder and Stoughton, 1964).

3. . . . forced to capitulate." From the Lisbon Report, quoted by Brown, *op. cit.*, p. 364.

4. . . . battle of Normandy." *Ibid.*, p. 721.

5. Eisenhower quickly decreed . . . *et seq.* Directive to Air Marshal Tedder, 18 June 1944. *The Papers of Dwight D. Eisenhower—the War Years*, ed. by A. D. Chandler, Vol. 3, note #1758.

6. . . . German scientist." Brown, *op. cit.*, p. 198.

7. . . . details are inaccurate." Churchill, *Second World War*, Vol. 5, pp. 226–227.

8. . . . fit in with these." Babington-Smith, *op. cit.*, p. 212.

9. . . . in position for launching." *Ibid.*, p. 229.

10. . . . came to hate them." Interview with Alvin Katz.

11. . . . to May 1941." Collier, *op. cit.*, p. 46.

12. . . . abandon the rest." Brookes, *op. cit.*, p. 198.

13. . . . named Wernher von Braun. For a general comprehensive

survey of von Braun's career, see *The New York Times*, 18 June 1977, pp. 1, 24.

14. . . . overgrown track itself." Babington-Smith, *op. cit.*, p. 236.

15. . . . more or less fortuitously." Collier, *op. cit.*, pp. 115–116.

16. . . . were completely mobile." Babington-Smith, *op. cit.*, pp. 236–237.

17. . . . and the guns sixty-five." Churchill, *Second World War*, Vol. 6, p. 48.

18. Hitler's bloody appetite . . . *et seq.* Allied V-weapon casualty figures vary somewhat from account to account. I have quoted, as probably the most accurate, Churchill's statistics. *Ibid.*, Vol. 6, pp. 48–59, 52–53, 55.

19. . . . have been written off." Eisenhower, *op. cit.*, p. 260.

20. . . . what retaliation is!" Albert Speer, *Spandau, the Secret Diaries* (New York: Macmillan, 1976), Pocket Book edition, p. 49.

11
SORCERY IN NORTH AFRICA

1. . . . and great general." Churchill, *Second World War*, Vol. 3, p. 200.

2. . . . bogey-man to our troops," Young, *op. cit.*, p. 12.

3. . . . dedicated and apolitical." Brown, *op. cit.*, p. 93.

4. . . . a non-pareil." Young, *op. cit.*, p. 13.

5. . . . demented seismograph." *Ibid.*, p. 79.

6. . . . the red flame." Jasper Maskelyne, *Magic—Top Secret* (London: Stanley Paul and Co., 1949), p. 47.

7. . . . these crucial targets." Barkas, *op. cit.*, p. 123.

8. . . . their concealed hangars." *Ibid.*, p. 125.

9. "A few weeks . . . *et seq.* Interview by David Fisher with Stephen Sykes.

10. . . . have won the battle." Young, *op. cit.*, p. 103. *Cf.* Churchill, *Second World War*, Vol. 3, pp. 565–566.

11. . . . in a five-ton truck." Maskelyne, *op. cit.*, p. 59.

12. . . . BE TAKING PLACE. *Ibid.*, p. 63.

13. . . . to turn turtle." *Ibid.*, p. 64.

14. . . . for open water." *Ibid.*, p. 64.

15. . . . such deadly meaning." *Ibid.*, p. 178.

16. . . . up to that time." Barkas, *op. cit.*, p. 188.

12
BAIT FOR THE FOX

1. . . . shipping was destroyed." Churchill, *Second World War*, Vol. 4, p. 588.

2. . . . known as the 'Kondor' mission." For a full detailing of this espionage operation see Brown, *op. cit.*, pp. 104–111.

3. . . . taking back its wounded." Paul Carell, *The Foxes of the Desert* (New York: E. P. Dutton, 1961), Bantam edition, p. 249.

4. . . . bloody well got to!" Brown, *op. cit.*, p. 118.

5. . . . as seen from the air." Barkas, *op. cit.*, p. 198.

6. . . . before dawn." *Ibid.*, p. 203.

7. . . . was already lost." Young, *op. cit.*, p. 170.

8. . . . was to be assaulted." Churchill in House of Commons, 11 November 1942, per Barkas, *op. cit.*, pp. 215–216.

9. . . . lower price in blood." *Ibid.*, p. 216.

13
MINCEMEAT SWALLOWED WHOLE

1. The top secret dispatch . . . *et seq.* Ewen Montagu, *The Man Who Never Was* (Philadelphia and New York: J.B. Lippincott, 1954), p. 117. Details in this chapter re "Major Martin" are based largely on Montagu's own story, to which I am indebted, and to which I have appended relevant facts and material as indicated.

2. is the inscription . . . *et seq.* Epitaph, *ibid.*, p. 116.

3. . . . North African shores." Churchill, *Second World War*, Vol. 4, p. 780.

4. . . . know it was Sicily." Montagu, *op. cit.*, p. 24.

5. . . . suddenly dropped." *Ibid.*, p. 25.

6. . . . or at least copies." *Ibid.*, p. 25.

7. . . . protect our interests." *Ibid.*, p. 78.

8. He might bring . . ." *et seq. Ibid.*, p. 65.

9. . . . had gone to war." *Ibid.*, p. 109.

10. . . . eight days before." *Ibid.*, p. 114.

11. . . . to play its part." *Ibid.*, p. 115.

12. . . . now very comfortable." *Ibid.*, p. 117.

13. . . . history of the world." Ernie Pyle, *Brave Men* (New York: Grosset and Dunlap, 1943), p. 8.

14. . . . we would select." Eisenhower, *op. cit.*, p. 174.

15. . . . above suspicion." Montagu, *op. cit.*, p. 129.

16. . . . points to Sardinia." *Ibid.*, p. 141.

17. But at this point . . . *et seq.* Brown, *op. cit.*, pp. 610–612.

18. The Wilhelmina anecdote is from James Leasor's *The Clock With Four Hands*, based on the memoirs of General Sir Leslie Hollis (New York: Reynal, 1959), p. 254.

19. . . . reveal their presence." John Toland, *Battle—the Story of the Bulge* (New York: Random House, 1959), p. 21.

20. . . . a remarkable achievement." Montagu, *op. cit.*, p. 12.

21. . . . and American lives." *Ibid.*, p. 17.

22. . . . lifetime of service." *Ibid.*, p. 17.

14
THE CONJURY OF FANATICISM

1. . . . losses sustained there." Abraham Rothberg, *Eyewitness History of World War II* (New York: Bantam Books, 1962), Vol. 3, p. 185.

2. . . . artificial construction." Report on Japanese camouflage, Hq. AAF, SWPA, dated 25 July 1943, p. 6. Many ruses and decoys described in this chapter are documented in the above report and in WD pamphlet 20–21, dated 12 Sept. 1944; also in British PRO file #40/2198 (AIR).

3. . . . 78 bullet holes." *Intelligence Digest* (Blacksmith, Oct. 1943), p. 65.

4. . . . problem in assault." *et seq.* Baldwin, *Battles Lost and Won*, p. 316.

5. . . . seemed on fire." Miller's account of the Eniwetok landings appears in *The Best from Yank* (New York: E. P. Dutton, 1945), "Surprise Party at Eniwetok," pp. 250–252.

6. . . . tunnels and corridors." Rothberg, *op. cit.*, Vol. 4, p. 176.

7. . . . were there waiting." *et seq.* Interview with Irwin Nelson.

8. and in 1946. . . *et seq.* U.S. Strategic Bombing Survey (Pacific), 1946—PI Section, p. 5.16.

9. . . . major importance." *Ibid.*, p. 5.11.

10. . . . a single system." *Ibid.*, p. 5.13.

11. While Army Intelligence . . . *et seq.* Baldwin, *Battles Lost and Won*, footnote #7, p. 629.

15
NAVIES IN MOTLEY

1. . . . British naval nerves." Tuchman, *op. cit.*, p. 326.

2. Erudite John Kerr . . . *et seq. Britannica, op. cit.*, Vol. 13, p. 316.

3. . . . blue, gray, and green. Sample Camouflage Design Sheets, USN regulations, and descriptions of various paint patterns can be found in *USN Camouflage, WW II* by L. Sowinski and T. Walkowiak (Philadelphia: The Floating Drydock, 1976).

4. We were scheduled . . ." *et seq.* Interview with Robert Paulus.

5. . . . after leaving harbor." Ian V. Hogg and J. B. King, *German and Allied Secret Weapons of WW II* (New Jersey: Chartwell, 1976), pp. 53–54.

6. In the long history . . . *et seq.* For a complete account of the operations of the *Atlantis* and other German disguised merchantmen, see David Woodward's *The Secret Raiders* (London: New English Library, 1975). Cf. *Secrets and Spies*, op. cit., pp. 138–144.

7. . . . naval and air force." Edwin P. Hoyt, *How They Won the War in the Pacific* (New York: Weybright and Talley, 1970), pp. 489–490.

8. . . . of any kind." Bombing Survey, *op. cit.*, p. 5.5.

9. . . . that Oahu was." David Kahn, *The Code-Breakers* (New York: Macmillan, 1967), New American Library edition, p. 307.

10. . . . the whole Pacific war." *Ibid.*, p. 309.

11. . . . out of place." *Ibid.*, p. 314.

16
COAL DUST, SMOKE, AND SILVER FOIL

1. . . . lost their significance." Gerald Pawle, *The Secret War* (New York: William Sloane, 1957), p. 187.

2. . . . from the Trent Valley." *Ibid.*, pp. 187–188.

3. Delighted with the . . . *et seq. Ibid.*, pp. 189–190.

4. . . . doing it with." Pyle, *Here Is Your War*, pp. 137–138.

5. . . . in World War II." Leo P. Brophy, Wyndham D. Miles, and Rexmond C. Cochrane, *U.S. Army in World War II—The Chemical Warfare Service: From Laboratory to Field* (Washington, D.C.: OMH, 1959), p. 197. Additional facts re wartime uses of smoke screening, pp. 197–219.

6. One late model . . . *et seq. Intelligence Digest, op. cit.*, p. 31.

7. . . . eastern tributaries." Eisenhower, *op. cit.*, p. 388.

8. To increase the . . . *et seq.* A detailed record of this cover and deception plan can be found in *Conquer—The Story of the Ninth Army* (Washington, D.C.: Infantry Journal Press, 1947), pp. 216–219.

9. . . . during the war." Eisenhower, *op. cit.*, p. 390.

10. . . . in cloud and fog." Churchill, *Second World War*, Vol. 2, p. 383.

11. . . . were two chickens." *Ibid.*, Vol. 3, p. 46.

12. . . . the deception thrust." Brown, *op. cit.*, p. 525.

17
PHANTASM AND REALITY: AN EPILOGUE

1. The illusion was . . . *et seq.* This noted illusion has been well described by Houdini's biographer, the late William L. Gresham, in *Houdini* (New York: Henry Holt, 1959), pp. 195–198.

2. Most important of all . . . *et seq.* For details re the U.S. SAMOS system and the Soviets' COSMOS satellites, see Brookes, *op. cit.*, pp. 227–231.

3. . . . intelligence organizations." Speer, *Inside the Third Reich*, p. 455.

4. . . . intelligence in general." *Ibid.*, p. 455.

5. . . . the other is incomplete." Baldwin, *Battles Lost and Won*, Introduction, page xii.

(In addition to the foregoing sources and other military histories and records, the following U.S. Army field manuals also provided much help: FM 5-20, FM 5-20B, FM 5-20E, FM 5-20G, FM 5-20H, FM 5-23, FM 30-21, FM 11-151 and TM 5-267.)

Index